The Busy Librarian's Guide to Information Literacy
in Science and Engineering

Edited by
- Katherine O'Clair
- Jeanne R. Davidson

Association of College and Research Libraries
A Division of the American Library Association
Chicago 2012

The paper used in this publication meets the minimum requirements of American National Standard for Information Sciences–Permanence of Paper for Printed Library Materials, ANSI Z39.48-1992. ∞

Library of Congress Cataloging-in-Publication Data

The busy librarians guide to information literacy in science and engineering / edited by Katherine O'Clair and Jeanne Davidson.
    pages cm
 Includes bibliographical references.
 ISBN 978-0-8389-8619-6 (pbk.)
 1. Science--Study and teaching (Higher) 2. Engineering--Study and teaching (Higher) 3. Science--Information resources--Study and teaching (Higher) 4. Engineering--Information resources--Study and teaching (Higher) 5. Information literacy--Study and teaching (Higher) 6. Information literacy--Standards. 7. Communication in science. 8. Communication of technical information. 9. Science and technology libraries--Reference services. 10. Academic libraries--Relations with faculty and curriculum. I. O'Clair, Katherine. II. Davidson, Jeanne.
 Q181.B968 2012
 507.1'1--dc23

                                        2012028043

Printed in the United States of America.

16 15 14 13 12   5 4 3 2 1

TABLE OF CONTENTS

INTRODUCTION

Katherine O'Clair
Agriculture and Environmental Sciences Librarian
California Polytechnic State University

Jeanne R. Davidson
Head, Academic Program Services
Arizona State University

HOW TO USE THIS BOOK

> "The landscape for scientific information is almost incomprehensibly complex. The literature is both deep in content and broad in scope, with a phenomenal amount of published scientific information. There are thousands of journals, hundreds of thousands of books, millions of articles and millions more references." (*Research Within the Disciplines*, p. 173)

Information literacy focuses on an individual's ability to effectively identify an information need, navigate the information universe, find what is needed, and synthesize it into new knowledge. The information literate individual will repeat this process continually throughout his or her lifetime as new information needs arise. As mentioned in the above quote, in the sciences (as well as engineering and technology) the variety, size and scope of the information landscape requires the information literate individual to possess specific knowledge and employ specialized strategies to be competent and successful. Our professional engagement in information literacy in science and related disciplines comes from our strong interest and background in the sciences and from many years of educating and working with individuals to help them understand the complexities of finding, using, and synthesizing scientific or technical data and information. The library literature contains many articles on varying aspects of this topic, but very few, if any, books contribute to the discussion. This book is a representation of our interest and dedication to this important aspect of librarianship.

Information literacy in science, engineering and technology can be exciting and rewarding for the librarian who uses effective strategies and approaches to work with faculty and students in these disciplines. The purpose of this book is to provide all librarians with meaningful insight into information literacy in selected disciplines of science, engineering and technology. Each chapter supplies the characteristics and unique aspects of the discipline or situation; strategies and approaches for integrating information literacy through active learning along with associated challenges and opportunities; and important information about the discipline or situation that will be helpful to the librarian without formal education or experience in science, engineering or technology. In addition, each chapter identifies the ALA/ACRL/STS Task Force on Information Literacy for Science and Technology [STS-TFILST] (2006) Information Literacy Standards for Science and Engineering/Technology (hereafter, Standards) that are most relevant and important to that discipline or situation.

One of the many motivations for this book is the fact that librarians with responsibilities in science, engineering and technology librarianship may begin without experience or formal training in these disciplines. While education and experience in these fields is helpful in many ways, it is not required to be a successful subject librarian. With dedication and through practice, one can gain the unique skills that are needed for serving users in these disciplines, and this has been demonstrated over the years by many librarians who have done just that.

We have chosen to focus on those disciplines in which information literacy has been widely integrated into the curriculum. Many of the concepts and strategies discussed throughout the book can be applied to other disciplines that are not covered in this text. Also included are several chapters that address special formats or institutional situations in which science, engineering and technology information literacy is relevant. This book is designed to provide practical information and strategies that can be used immediately by both subject specialists and those with instructional responsibilities in science, engineering and/or technology. Readers may use selected content on an as-needed basis or read this book cover-to-cover and refer back to it when necessary. It is organized to allow either approach to be equally effective.

This introduction continues with an overview of information literacy in science, engineering and related disciplines. It also contains information about the Standards, their history and evolution, and how they can be applied in practice. The remainder of the book focuses on information literacy in the specific discipline or special cases. Each chapter includes a list of standards and performance indicators of particular importance along with discussion of key characteristics for the discipline or special case. Strategies for successfully interacting with faculty and/or students in the discipline are presented along with suggestions for librarians who may come to a discipline with no background in the field.

Chapters 2 through 5 address several disciplines in which information literacy is often integrated: engineering, chemistry, life and health sciences, and human nutrition. Chapter 2 addresses engineering as a discipline as well as presenting options for active learning and assessment that may be adapted to other disciplines. Chapter 3 introduces librarianship in both the life sciences and health sciences. Chemistry librarians have additional standards and have shared strategies in the literature. Chapter 4 provides a comprehensive overview of chemistry and additional disciplinary standards as well as a guide to some of the most useful strategies previously published. Chapter 5 discusses human nutrition as a scientific discipline and includes important information on dietetics standards along with the Standards addressed in this book.

Chapters 6 through 8 are special cases: patents, interdisciplinary science courses, and community colleges. Patents (Chapter 6) are an important information source for many disciplines including chemistry, engineering, computer science, and even business, yet they are often underutilized as a result of the difficulty associated with locating them and using the information contained within them. Interdisciplinary courses are becoming increasingly common at many institutions. Chapter 7 presents strategies used in a Remote Sensing course that pulls on several science fields and illustrates an ideal integration of information literacy into a single course. Many community colleges have science programs that have unique needs because of the nature of the institution. Chapter 8 contains a description of one such program with suggestions for how information literacy can be incorporated into the curriculum.

INFORMATION LITERACY IN SCIENCE, ENGINEERING AND TECHNOLOGY

Individuals who are information literate are able to fully engage in lifelong learning, because they can locate, evaluate, and use the information needed for any situation (American Library Association (ALA), 1989). Information literacy in higher education is fundamental because "developing lifelong learners is central to the mission of higher education institutions" (Association of College and Research Libraries (ACRL), 2000). Nearly all of the agencies that accredit colleges and universities in the United States include some aspect of information literacy in their curriculum criteria (Saunders, 2007). In addition, many disciplinary accreditation associations in science, engineering and technology (see Manuel, 2004) identify the components of information literacy in their curricular requirements, although they may not be specifically defined as such. The Information Literacy Competency Standards for Higher Education were developed as a way to assess students' skills and abilities with respect to information literacy (ACRL, 2000), and they have made an important and significant impact on the way information literacy is taught and assessed at colleges and universities around the world. Since their release in 2000, they have been the basis for the development of additional discipline-specific information literacy standards, including the Information Literacy Standards for Science and Engineering/Technology, which are the focus of this book.

The science, engineering, and technology disciplines share many of the general principles of information literacy with the humanities, social sciences, and other fields. These include the ability to determine the information need, access the information needed, evaluate it, incorporate and use the information, and understand the economic, legal and social issues associated with these activities (ACRL, 2000). Yet, disciplines in science, engineering and technology have some inherent differences that have implications for information literacy, such as the nature of the literature that is used for scholarly communication and the structure of the curriculum in these fields. In addition, students in these disciplines need to be well prepared for lifelong learning throughout their future professional careers, as many will operate daily in potentially life or death situations (i.e., practicing medicine, designing and constructing building structures, etc.). Consequently, information literacy in the sciences and engineering and technology fields deserves in depth attention.

Sci/Tech Scholarly Communication

The nature of scholarly communication in any discipline influences the paths used for seeking information and the sources an individual will consult. The science and related disciplines are no exception. Researchers and professionals in science, engineering and technology disciplines rely predominantly on non-monographic sources (e.g. articles, conference proceedings, technical reports, etc.) for communication in these fast-paced and constantly changing environments because they tend to be more specific and immediate than books. The scholarly peer-reviewed article is the predominant form of scholarly communication in these disciplines, and consequently, most professors and instructors will want their students to find, use, synthesize and cite information from these types of sources. Subscription and renewal fees for scholarly, peer-reviewed journals in the science, technology, and engineering disciplines tend to be quite high, and few libraries have the luxury of subscribing to all the discipline-specific journals faculty and students encounter in their information seeking activities. As a result of high subscription and renewal fees, many academic libraries have moved away from the "just in case" model for providing journals to a "just in time" strategy, in which document delivery and interlibrary loan provide the most cost effective way to provide access to the articles faculty and student need.

In addition to the scholarly peer-reviewed article, there are a number of other types of literature that are relevant and useful in these disciplines, including patents (discussed in Chapter 6), standards (discussed in Chapter 2) and the gray literature. The gray literature generally includes documents, such as technical reports and conference proceedings, that are not commercially published. Despite being widely cited in the literature, these documents tend to be difficult to locate, because they are usually not indexed and collected in the manner that is typical of other publications (Mathews, 2004).

Researchers and professionals (as well as students) must utilize specific strategies to navigate the distinctive information landscape of the science, engineering and technology disciplines. Many students take the "one size fits all" approach to searching the literature and assume that what they learned about "library research" in an introductory university writing course (or even in high school) applies to any discipline. An information seeking strategy that may be effective in the humanities will not be nearly as effective in searching for science, engineering or technology information because of the vast differences in the ways practitioners in these disciplines communicate information, knowledge and research. There are a wide variety of specialized resources, such as interdisciplinary and discipline-specific indexing and abstracting databases, but many students and even some faculty are unaware of these important tools for searching the literature (Kieft, 2008). Moreover, because many students are warned by their instructors to avoid Internet sources, they may steer away from important and useful documents that are only available on the websites of government agencies or other legitimate entities (e.g. university research centers, professional associations, etc.) (Kieft, 2008).

Common Curricular Challenges

There has been a perception that the science disciplines have been unreceptive to information literacy, but Manuel (2004) argues that the evidence and research do not support this idea. The resistance to information literacy that librarians may face is often a consequence of the nature of the curriculum rather than the lack of interest on the part of faculty in those disciplines. It is important for librarians assigned to science, engineering, or technology areas to understand the curriculum in these disciplines, as it presents both challenges and opportunities for integrating information literacy.

Disciplines in science, engineering and technology utilize a combination of lecture and laboratory or activity courses to deliver the curricular content. The lecture often uses an instructional approach that is passive and instructor focused, while laboratory or activity courses allow students to learn in a much more active and engaged manner. In most lecture-based courses, the majority of the information needed for learning is provided by the professor during the lecture or through assigned textbook readings. Many of the concepts taught have been known for many years, and the associated information is difficult to glean from the recent literature. These courses typically assess student learning through midterm and final exams. Thus, students have little need to consult additional information sources, unless a term paper or research project is assigned.

Laboratory or activity courses often taught in conjunction with the lecture course offer more opportunities for integrating information literacy. The typical assessment methods used for these include frequent quizzes and laboratory reports along with a culminating practicum exam. Many instructors assign written lab reports in which they require students to consult and cite the literature, so they may be very open to the idea of utilizing the librarian to teach students how to search, use and cite the literature. This often presents the opportunity for the librarian to work directly with students and connect course-specific content with information literacy concepts. Regardless of the format of the course, it is important to note that most courses already have more content than there is time to cover, so instructors may be reluctant, or even unwilling, to sacrifice valuable instructional time for the sake of adding some form of information literacy or library instruction. This circumstance is not insurmountable, but it often requires working closely with the instructor to negotiate when and how to incorporate what is often perceived as extraneous content.

Instructional Strategies
Another important consideration involves the instructional strategies used to teach information literacy. Leonard (1997) stresses the importance of actively engaging students, particularly those in the sciences, and recommends that "an active learning environment is the most appropriate mode of instruction" (pg. 16). Because of their predominant learning styles, students in science, engineering or technology disciplines will prefer and be more receptive to active learning approaches for teaching information literacy. Several publications (see Additional Resources) offer numerous examples of strategies for integrating active learning into library and information literacy instruction. Although these do not focus specifically on science, engineering, and technology, they include several chapters that address these disciplines and many of the strategies and activities can be adapted for science, engineering and technology.

Both the Science and Technology Section (STS) and the Instruction Section (IS) of the Association of College and Research Libraries (ACRL) provide wikis that contain relevant and useful information about instructional strategies for information literacy in specific disciplines. The Science Information Literacy wiki, which was created by the STS Information Literacy Committee, offers a variety of useful materials including teaching tips, subject guides and assessment resources. Librarians responsible for information literacy instruction in science and related disciplines are encouraged to contribute their ideas and share their strategies on this wiki. In addition to the wiki, the STS Information Literacy

Committee hosts monthly chat session using social media to discuss issues and concerns related to information literacy in science, engineering and technology. The IS Information Literacy in the Disciplines wiki provides citations and links for resources that discuss discipline-specific information literacy strategies and approaches, including those for the science and engineering disciplines. Practitioners are encouraged to submit recommendations, ideas and strategies to the IS Information Literacy in the Disciplines Committee for inclusion in the wiki.

Assessment

Assessment is an important part of information literacy instruction as it provides feedback (sometimes immediate) on the learning process for both the student and the instructor. For the student, assessment provides feedback on how well he or she is understanding and learning the material. From the instructor's perspective, the assessment shows how effective he or she has been at conveying the information and knowledge, and it also helps to identify areas where improvements need to be made (Burkhardt, MacDonald, & Rathemacher, 2010). Assessment has many forms and functions. Angelo and Cross (1993) outline numerous classroom assessment techniques (CATS) that can be used to gauge student learning in the classroom. They recommend utilizing these techniques early on and throughout the instruction, rather than at the end, so that any gaps can be addressed immediately. This strategy helps both instructors and students improve their teaching and learning (Angelo & Cross, 1993). Walsh (2009) identifies a variety of methods that can be used to assess information literacy including self-assessments, quizzes or tests, analysis of bibliographies. Chapter 2 of this book presents additional information about assessment with specific strategies for assessing information literacy in engineering, many of which can be applied to a variety of other disciplines in science, engineering and technology.

INFORMATION LITERACY STANDARDS FOR SCIENCE AND ENGINEERING/TECHNOLOGY

The Information Literacy Standards for Science and Engineering/Technology (STS-TFILST, 2006) were the first disciplinary standards approved as a supplement to the Information Literacy Competency Standards for Higher Education (ACRL, 2000). They are intended to apply to all areas of science and engineering/technology. The field of mathematics is not addressed in these standards.

The Science & Technology Section (STS) of the Association of College and Research Libraries (ACRL) developed the Standards, which are drawn from the Information Literacy Competency Standards for Higher Education (ACRL, 2000) with modifications based on the nature of the scientific and technical literature and publishing practices. Their development involved a multi-year process of drafting, soliciting feedback from science and engineering librarians in multiple disciplines, and revising based on the suggestions and comments received. Although the Standards were created by STS, librarians in other organizations, such as the Special Libraries Association (SLA) and the Engineering Libraries Division (ELD) of the American Society of Engineering Education (ASEE), were consulted on the concepts and content as the work progressed. This ensured that all areas of science and engineering were considered as the standards developed. The Standards then passed through an approval process that involved librarians with expertise in instruction and pedagogy as well as ACRL's Standards and Accreditation Committee, which oversees all standards or guidelines approved by ACRL.

Given the ever-changing nature of science and engineering, the purpose of the standards is to establish behavioral competencies to enable the graduation of life-long learners into the ranks of practicing scientists and engineers. The Standards are articulated in five standards and twenty five performance indicators, each with accompanying outcomes (STS-TFILST, 2006). As with all ACRL standards and guidelines, the Standards are required to undergo review every five years. They were reviewed in 2010–11 and are currently in the process of being revised. Although at this time the extent of the revisions is unknown, the basic framework will remain largely unchanged. Proposed revisions are needed to ensure they remain up to date with current scholarly communications models in science and engineering/technology.

The Standards can be used for a variety of purposes and at several levels. Used at the standard and/or performance indicator level, they serve as a springboard for conversation with disciplinary faculty about skill sets that their students will need as researchers or practicing scientists or engineers. The outcomes attached to each performance indicator provide a level for focusing student learning and for assessing student achievement and may be used in conjunction with college or departmental learning outcomes. Because information literacy instruction is often conducted in single, stand-alone sessions, prioritizing learning outcomes prior to the instruction session is critical so students will not be overwhelmed with too much information.

The Standards are broadly applicable in science and engineering/technology. However, several disciplines have standards generally applicable to information literacy set by professional organizations in the specific fields. Examples include engineering (ABET, discussed in Chapter 2), chemistry (American Chemical Society, discussed in chapter 4) and human nutrition (Academy of Nutrition and Dietetics, discussed in chapter 5). Identifying existing disciplinary standards and relating them to the Standards is a worthwhile investment of time. Through the process, the librarian gains a more detailed understanding of the learning outcomes faculty are working to achieve. This knowledge can lead to conversations between faculty and librarians that uncover the common ground where collaboration with a librarian adds value to student learning.

ADDITIONAL RESOURCES

ACRL Instruction Section Information Literacy in the Disciplines Committee. (22 November 2011). Information literacy in the disciplines. Available at http://wikis.ala.org/acrl/index.php/Information_literacy_in_the_disciplines

ACRL Science and Technology Section Information Literacy Committee. (2011). Science information literacy. Available at http://wikis.ala.org/acrl/index.php/Science_Information_Literacy

Angelo, T. A. & Cross, K. P. (1993). *Classroom assessment techniques: A handbook for college teachers* (2nd ed.). San Francisco: Jossey-Bass.

Burkhardt, J. M., & MacDonald, M. C. (2010). *Teaching information literacy: 50 standards-based exercises for college students* (2nd ed.). Chicago: ALA.

Gradowski, G., Snavely, L., & Dempsey, P. (Eds). (1998). *Designs for active learning: A sourcebook of classroom strategies for information education.* Chicago: ACRL.

Sittler, R., & Cook, D. (Eds.). (2009). *The library instruction cookbook: 50+ active recipes for 1-shot sessions.* Chicago: ACRL.

REFERENCES

American Library Association (ALA). (1989). Presidential Committee on Information Literacy: Final report. Retrieved from http://www.ala.org/ala/acrl/acrlpubs/whitepapers/presidential.htm

Association of College and Research Libraries (ACRL). (2000). Information literacy competency standards for higher education. Retrieved from http://www.ala.org/acrl/sites/ala.org.acrl/files/content/standards/standards.pdf.

ALA/ACRL Science and Technology Section (STS) Task Force on Information Literacy for Science and Technology [STS-TFILST]. (2006). Information literacy standards for science and engineering/technology. Retrieved from http://www.ala.org/acrl/standards/infolitscitech

Angelo, T. A. & Cross, K. P. (1993). *Classroom assessment techniques: A handbook for college teachers* (2nd ed.). San Francisco: Jossey-Bass.

Burkhardt, J. M., MacDonald, M. C., & Rathemacher, A. J. (2010). *Teaching information literacy: 50 standards-based exercises for college students.* Chicago: American Library Association.

Kieft, R. H. (Ed.) (2008). *Guide to reference* [electronic format]. Chicago: ALA. Retrieved from http://library.lib.asu.edu/record=b5467359~S3

Kraus, J. R. (2007). Research in the sciences. In P. Keeran, S. Moulton-Gertig, M. Levine-Clark, N. Schlotzhauer, E. Gil, C. Brown, J. Bowers (Eds.), *Research within the disciplines: Foundations for reference and library instruction* (pp. 173–200). Lanham, MD: Scarecrow Press.

Leonard, W. H. (1997). How do college students learn science? In E. Siebert, M. Caprio, & C. Lyda (Eds.), *Methods of effective teaching and course management for university and college science teachers* (pp. 5–20). Dubuque, IA: Kendall/Hunt.

Manuel, K. (2004). Generic and discipline-specific information literacy competencies: The case of the sciences. *Science & Technology Libraries, 24,* 279–308.

Mathews, B. S. (2004). Gray literature. *College & Research Libraries News, 65,* 125–128.

Saunders, L. (2007). Regional accreditation organizations' treatment of information literacy: Definitions, collaboration, and assessment. *Journal of Academic Librarianship, 33,* 317–326.

Walsh, A. (2009). Information literacy assessment: Where do we start? *Journal of Librarianship and Information Science, 41,* 19–28.

Corresponding Chapters for the Information Literacy Standards for Science and Engineering/Technology Addressed in this Book

Standard 1: The information literate student determines the nature and extent of the information needed.	
1.1 Defines and articulates the need for information.	Engineering Interdisciplinary Science
1.2 Identifies a variety of types and formats of potential sources for information.	Engineering Chemistry Interdisciplinary Science Community Colleges
1.3 Identifies a variety of types and formats of potential sources for information.	Chemistry Life and Health Science Patents Interdisciplinary Science Community Colleges
1.4 Considers the costs and benefits of acquiring the needed information.	Engineering Chemistry Patents
Standard 2: The information literate student acquires needed information effectively and efficiently.	
2.1 Selects the most appropriate investigative methods or information retrieval systems for accessing the needed information.	Engineering Chemistry Life and Health Science Human Nutrition Interdisciplinary Science
2.2 Constructs and implements effectively designed search strategies.	Engineering Chemistry Life and Health Science Human Nutrition Patents Interdisciplinary Science Community Colleges
2.3 Retrieves information using a variety of methods.	Chemistry Human Nutrition
2.4 Refines the search strategy if necessary.	Chemistry Human Nutrition
2.5 Extracts, records, transfers, and manages the information and its sources.	Chemistry Human Nutrition Interdisciplinary Science

Standard 3: The information literate student critically evaluates the procured information and its sources, and as a result, decides whether or not to modify the initial query and/or seek additional sources and whether to develop a new research process.	
3.1 Summarizes the main ideas to be extracted from the information gathered.	Chemistry Interdisciplinary Science
3.2 Selects information by articulating and applying criteria for evaluating both the information and its sources.	Engineering Chemistry Life and Health Science Human Nutrition Patents Interdisciplinary Science Community Colleges
3.3 Synthesizes main ideas to construct new concepts.	Chemistry Human Nutrition
3.4 Compares new knowledge with prior knowledge to determine the value added, contradictions, or other unique characteristics of the information.	Engineering Chemistry Human Nutrition Interdisciplinary Science
3.5 Validates understanding and interpretation of the information through discourse with other individuals, small groups or teams, subject-area experts, and/or practitioners.	Engineering Chemistry Human Nutrition
3.6 Determines whether the initial query should be revised.	Engineering Patents Interdisciplinary Science
3.7 Evaluates the procured information and the entire process.	Engineering Human Nutrition Interdisciplinary Science
Standard 4: The information literate student understands the economic, ethical, legal, and social issues surrounding the use of information and its technologies and either as an individual or as a member of a group, uses information effectively, ethically, and legally to accomplish a specific purpose.	
4.1 Understands many of the ethical, legal and socio-economic issues surrounding information and information technology.	Engineering Chemistry Human Nutrition Patents
4.2 Follows laws, regulations, institutional policies, and etiquette related to the access and use of information resources.	Engineering Chemistry Interdisciplinary Science
4.3 Acknowledges the use of information sources in communicating the product or performance.	Engineering, Chemistry Human Nutrition Interdisciplinary Science

4.4 Applies creativity in use of the information for a particular product or performance.	Engineering Chemistry
4.5 Evaluates the final product or performance and revises the development process used as necessary.	Engineering Chemistry
4.6 Communicates the product or performance effectively to others.	Engineering Chemistry Human Nutrition
Standard 5: The information literate student understands that information literacy is an ongoing process and an important component of lifelong learning and recognizes the need to keep current regarding new developments in his or her field.	
5.1 Recognizes the value of ongoing assimilation and preservation of knowledge in the field.	Chemistry Patents Community Colleges
5.2 Uses a variety of methods and emerging technologies for keeping current in the field.	Engineering Chemistry Life and Health Science

IMPORTANT INFORMATION LITERACY STANDARDS FOR ENGINEERING

1. The information literate student determines the nature and extent of the information needed.
    1. Defines and articulates the need for information.
        a. Identifies and/or paraphrases a research topic, or other information need such as that resulting from an assigned lab exercise or project.
        c. Develops a hypothesis or thesis statement and formulates questions based on the information need.
    2. Identifies a variety of types and formats of potential sources for information.
        c. Identifies the value and differences of potential resources in a variety of formats (e.g., multimedia, database, website, data set, patent, Geographic Information Systems, 3-D technology, open file report, audio/visual, book, graph, map).
    4. Considers the costs and benefits of acquiring the needed information.
        c. Formulates a realistic overall plan and timeline to acquire the needed information.

2. The information literate student acquires needed information effectively and efficiently.
    1. Selects the most appropriate investigative methods or information retrieval systems for accessing the needed information.
    2. Constructs and implements effectively designed search strategies.

3. The information literate student critically evaluates the procured information and its sources, and as a result, decides whether or not to modify the initial query and/or seek additional sources and whether to develop a new research process.
    2. Selects information by articulating and applying criteria for evaluating both the information and its sources.
    4. Compares new knowledge with prior knowledge to determine the value added, contradictions, or other unique characteristics of the information.
        a. Determines whether information satisfies the research or other information need.
        e. Determines probable accuracy by questioning the source of the information, limitations of the information gathering tools or strategies, and the reasonableness of the conclusions.

IMPORTANT INFORMATION LITERACY STANDARDS FOR ENGINEERING

  f. Integrates new information with previous information or knowledge.

 5. Validates understanding and interpretation of the information through discourse with other individuals, small groups or teams, subject-area experts, and/or practitioners.

  b. Works effectively in small groups or teams.

 6. Determines whether the initial query should be revised.

 7. Evaluates the procured information and the entire process.

4. The information literate student understands the economic, ethical, legal, and social issues surrounding the use of information and its technologies and either as an individual or as a member of a group, uses information effectively, ethically, and legally to accomplish a specific purpose.

 1. Understands many of the ethical, legal and socio-economic issues surrounding information and information technology.

  d. Demonstrates an understanding of intellectual property, copyright, and fair use of copyrighted material and research data.

 2. Follows laws, regulations, institutional policies, and etiquette related to the access and use of information resources.

  f. Demonstrates an understanding of what constitutes plagiarism and does not represent work attributable to others as his/her own. This includes the work of other members of research teams.

 3. Acknowledges the use of information sources in communicating the product or performance.

  a. Selects an appropriate documentation style for each research project and uses it consistently to cite sources.

 4. Applies creativity in use of the information for a particular product or performance.

 5. Evaluates the final product or performance and revises the development process used as necessary.

 6. Communicates the product or performance effectively to others.

5. The information literate student understands that information literacy is an ongoing process and an important component of lifelong learning and recognizes the need to keep current regarding new developments in his or her field.

 2. Uses a variety of methods and emerging technologies for keeping current in the field.

ENGINEERING

Sheila J. Young
Academic Professional Emerita
Arizona State University

INTRODUCTION

Engineering and technology is a broad subject area. In addition to information common across the subject area, each sub-discipline has its unique information needs. Engineers require information from a wide variety of sources including articles from journals and conference proceedings, data compilations, books, government documents, regulations, product information, technical reports, patents, standards, and business resources. Although the engineering curriculum is very full, it is important to demonstrate that information literacy skills are not an add-on but rather support the learning outcomes of a course and the program. McCullough (2006) emphasizes the importance of information literacy to the practice of engineering stating, "Information literacy is more than a necessary element of education: it is the very *essence* of education" (Abstract). Using the Columbia shuttle disaster of 2003 as an example, McCullough (2006) makes a strong case for engineers to possess and apply information literacy skills: "some of the causes cited [by NASA] as direct or indirect causes were incorrect evaluation of information, inappropriate use of data sources, and failure to seek additional information when necessary" (Do Engineers Need It Section). She lists five of the causes of the failure included in the NASA report and links each to the failure to apply one or more of the skills identified in the ALA/ACRL/STS Task Force on Information Literacy for Science and Technology [STS-TFILST] (2006) Information Literacy Standards for Science and Engineering/Technology (hereafter, Standards).

Incorporation of information literacy into engineering courses and the engineering curriculum has gained momentum in recent years. Although it is possible to teach information literacy as a separate course, the packed curriculum of engineering programs makes this approach impractical. Information literacy can be incorporated into existing courses as a stand-alone endeavor or as part of a more comprehensive plan to build sequenced information literacy instruction and activities into the curriculum.

This chapter addresses the relationship of the Standards to the ABET (formerly known as the Accreditation Board for Engineering and Technology) accreditation requirements and documents. It presents examples of the incorporation of information literacy instruction and activities into stand-alone courses as well as into courses within a planned progression of instruction throughout a curriculum. In addition, this chapter includes suggestions for developing and assessing learning modules.

ABET ACCREDITATION AND INFORMATION LITERACY

Engineering programs are accredited by ABET. The documents used in the accreditation process include the criteria (ABET Accreditation, 2011) for each program and self-study questionnaires and templates (ABET Self Study Questionnaire Templates, 2011). Because the curriculum must fulfill the requirements of the ABET criteria, alignment of information literacy in support of the criteria as described in the questionnaire creates an opportunity for librarians to contribute meaningfully to the educational process and simultaneously fulfill the requirements of ABET.

In the ABET Criteria and Self-Study Questionnaire Templates for Engineering, Computer Science, and Applied Science, library services are included in the section titled "Facilities."

> E. Library Services
> Describe and evaluate the capability of the library (or libraries) to serve the program including the adequacy of the library's technical collection relative to the needs of the program and the faculty, the adequacy of the process by which faculty may request the library to order books or subscriptions, the library's systems for locating and obtaining electronic information, and any other library services relevant to the needs of the program. (ABET Self-Study Questionnaire Templates, 2011, Facilities section).

Although ABET does not specifically mention information literacy, the relationship between the criteria and information literacy has focused on Criterion 3(a-k). ABET Engineering Criterion 3(i) states that the student will develop "recognition of the need for, and an ability to engage in life-long learning." (ABET Accreditation Criteria, 2011). Shuman, Besterfield-Sacre, and McGourty (2005) propose that lifelong learning includes:

> Demonstrate reading, writing, listening, and speaking skills; demonstrate an awareness of what needs to be learned; follow a learning plan; identify, retrieve and organize information; un-

derstand and remember new information; demonstrate critical thinking skills; and reflect on one's own understanding. (p. 49)

Although Shuman et al. (2005) was published before the approval of the Standards, it is clear they address these skills, and the outcomes provide evidence for fulfilling Criterion 3(i).

The maps of the Standards to the ABET criteria and the ABET Self-Study Questionnaire provide librarians a starting point for integrating information literacy into courses and the curriculum as part of the library services that support the program learning outcomes and the ABET criteria. Several authors have mapped ABET Criterion 3 to the Standards and include examples of assignments and/or assessments (McCullough 2006; Nelson & Fosmire 2010; Riley, Piccinino, Moriarity, and Jones, 2009). McCullough (2006) provides a map of all of the elements of the ABET 3(a-k) criteria to the Standards and describes assignments that develop information literacy skills in engineering students. Riley, Piccinino, Moriarty, and Jones (2009) describe the process of connecting Smith College's information literacy plan to ABET's criteria through mapping the Engineering program's outcomes to the ABET 3(a-k) outcomes as well as the Standards. They propose that ABET 3(b) ("an ability to design and conduct experiments, as well as to analyze and interpret data") be revised to "reflect importance of information literacy and the prominence of data derived from sources that are not original experiments conducted by a given engineer" and "support information literacy as a critical component of professional preparation for engineers" (Conclusion Section). To achieve this, they suggest that 3(b) could read "an ability to access and evaluate information as well as to design and conduct experiments to collect, analyze, and interpret original data" (Conclusion Section). Nelson and Fosmire (2010) discuss the process of curricular redesign to include information literacy in the evidence of accomplishing the ABET Criterion 3. They presented detailed tables mapping Criterion 3 (a-k) of the 2009–2010 ABET Criteria for Accrediting Engineering Technology Programs to specific performance indicators and outcomes of the Standards and listed examples of assessment strategies.

INTEGRATING INFORMATION LITERACY INTO COURSES AND THE CURRICULUM

The integration of information literacy varies from institution to institution. Based on the size of the institution and the number of librarians, institutions will have a variety of librarian to faculty and student ratios, so librarians need to be creative in designing the most effective means to meet the student and faculty needs. A librarian should begin by assessing the status of information literacy specific to his/her institution. Evaluate existing connections between the faculty and the library including what courses incorporate accessing, evaluating, and using information or data.

Librarians need to actively engage in outreach by identifying activities, events and organizations that offer entry points for developing relationships. Examples include orientations for new and transfer students, poster sessions, "Engineering Day" events, and engineering student organizations. Institution-specific programs, such as Arizona State University's Fulton Undergraduate Research Initiative (FURI) and Motivated Engineering Transfer Scholars (METS), also provide entry points for information literacy instruction for targeted groups. Librarians should also identify curricular entry points such as introductory and capstone courses, design classes, and specific courses that require students to search for data or information. In addition to faculty and graduate teaching assistants as people entry points, non-faculty academic advisors are a useful source of information about the students and the curriculum, and they can be helpful for making contact with teaching faculty as well. Librarians can discuss incorporating specific information skills and activities designed to support and facilitate the students accomplishing the goals and learning outcomes for a course or program.

Information literacy can be incorporated into a stand-alone course or into several courses as a unified progressive plan. A stand-alone session, by necessity, usually focuses on information literacy skills related to topic formation, specific resources, search strategies, and evaluation of retrieved information. More extensive integration facilitates a broader and higher level of information competency. For example, at Trinity University, information literacy was incorporated throughout the sequence of design courses (MacAlpine & Uddin, 2009). The content is shown in Table 2.1.

Table 2.1. Information Literacy in Design Courses at Trinity University	
Level	Topics and Activities for IL Instruction
Freshman	• Assessing information • Databases • Patents
Sophomore	• Patents • Product sources • Standards
Junior	• Safety requirements • Specifications • User manuals
Senior	• Individual Consultations
Adapted from (MacAlpine & Uddin, 2009)	

For curriculum development, Wiggins and McTighe (2005) provide a "backwards design" roadmap. This strategy involves three steps.

> "1. **Identify the desired result**. What should students know, understand, and be able to do? What is worthy of understanding? What enduring understandings are desired" (p. 17)?

> "2. **Determine acceptable evidence.** How will we know if students have achieved the desired results and met the standards? What will we accept as evidence of student understanding and proficiency" (p. 18)?

> "3. **Plan learning experiences and instruction**…What enabling knowledge (facts, concepts, principles) and skills (processes, procedures, strategies) will students need in order to perform effectively and achieve desired results? What activities will equip students with the needed knowledge and skills? What will need to be taught and coached and how should it best be taught, in light of performance goals? What materials and resources are best suited to accomplish these goals" (p. 18–19)?

These steps can be applied in developing a curriculum or an individual information literacy session.

THE DESIRED RESULT: WHAT DO I WANT STUDENTS TO KNOW AND BE ABLE TO DO?

Step one, the desired result, is expressed as learning outcomes. Using the Standards, one can generate learning outcomes important to engineering students. The broad outcomes in the Standards are that the student will be able to identify need, access, evaluate and use information, understand ethical and legal issues, and keep current. Welker, Fry, McCarthy, and Komlos (2010) provide details of a sequential information literacy program integrated into five courses. Using the ACRL Information Literacy Standards for Higher Education standards and outcomes, they developed 26 outcomes distributed among the sophomore, junior, and senior year. The following are selected examples of each level of what the students should be able to do.

Sophomore
1. Explore general information sources to increase familiarity with a topic,

2. Identify key concepts and terms that describe the information need,
3. Define a realistic overall plan and timeline to acquire the needed information
6. Evaluate a website for authority, reliability, credibility, purpose, viewpoint and suitability
18. Develop a thesis statement and formulate questions based on the information need.

Junior
19. Select efficient and effective approaches for accessing the information needed
21. Create a system for organizing the information

Senior
22. Recognize that existing information can be combined with original thought, experimentation, and/or analysis to produce new information
25. Extend initial synthesis, when possible, at a higher level of abstraction to construct new hypotheses that may require additional information

(Welker et al., 2010, Outcomes by Year section)

ASSESSMENT: WHAT IS ACCEPTABLE EVIDENCE?

The second step of backward design is to determine the evidence. How will I assess student learning? What is the evidence of learning? Evidence is often divided into direct (such as exams, portfolios, and performance evaluations) and indirect methods (such as surveys and questionnaires). ABET provides guidance on these two methods with a webinar and accompanying slides (Rogers, 2010). Another dimension of assessment is formative and summative ("Formative vs. Summative Assessment," n.d). Formative feedback assesses the progress of student learning and is used by instructors to improve teaching. Summative assessment is used to measure the level of competency of the student. Angelo and Cross (1993) developed classroom assessment techniques (CATS) as tools for formative assessment. These techniques provide very useful information to improve teaching. Examples of CATS using strategies from Angelo and Cross are included in Appendix 2.1. Summative assessment of student products or presentations can be graded using a rubric. ABET also provides guidance on the development of rubrics by means of a webinar and slides (Breidis, 2010).

Other helpful rubrics for assessing specific information literacy outcomes of engineering students are available. The "Information Literacy Rubric" in the Instruction Toolkit for Textiles and Engineering Services at North Carolina State University created by Nerz and McCord (2008) maps the learning outcomes of finding, locating, synthesizing, and presenting information to ABET Criterion 3 (b,c,e,I,k) and describes the skills at five levels of competency. The "Guide to Rating ABET Professional Skills" rubric created by the College of Engineering and Architecture and the Center for Teaching, Learning and Technology at Washington State University (2009) describes levels of competency for three information literacy outcomes for ABET 3(i): "identify need and determine how to obtain, access and evaluate sources, and analyze one's own biases and assumptions". Parker and Godavari (2011) created a detailed 5 scale rubric for finding, locating, analyzing, synthesizing and presenting information; connecting each category to the Canadian Engineering Accreditation Board (CEAB) criteria, the ABET criteria and the Standards.

Quizzes and tests are another method of summative assessment. It is challenging to develop a valid multiple choice test for information literacy. Wertz, Ross, Purzer, Fosmire, and Cardella (2011) at Purdue University have developed an alpha version of a multiple choice assessment tool. Students are presented with a memo and both multiple choice and open ended questions related to the content.

Table 2.2. Teaching Techniques for Specific Learning Styles	
Learning Style	Teaching suggestion
Active learners	provide in-class practice
Reflective learners	reflection exercises
Intuitive learners	teach basic principles and theories
Sensing learners	teach basic principles and theories in context of concrete applications and many application examples
Thinkers and sequential learners	present new material logically and sequentially
Feelers and Global learners	connect information to prior knowledge and global and social issues
All learners	Incorporate both visuals (pictures, diagrams, flow charts, concept maps, demonstrations etc. and verbal presentations spoken and written)
Source: Felder (2010) p.4	

LEARNING EXPERIENCES: WHAT ARE THE TEACHING STRATEGIES?

Understanding the learning styles of students is useful in developing effective teaching techniques. In their highly cited early paper on learning and teaching, Felder and Silverman (1988) describe the learning styles of engineering students as well as the teaching styles of faculty: "Many or most engineering students are visual, sensing, inductive, and active, and some of the most creative students are global" (p. 680). Learning styles and teaching effectiveness continue to be very active areas of research in engineering education and provide information for engineering librarians on how to teach information literacy to this group. In a recent discussion of learning styles and teaching Felder (2010) discusses the concepts as well as some teaching suggestions for a variety of learning styles (Table 2.2) and states, "Teaching to address all categories of a learning styles model is not a radical idea, and specific suggestions for how to do it should look familiar to anyone who has studied the literature of effective pedagogy" (p. 4).

An information literacy component can be incorporated into existing assignments by requiring the students to find the information to solve a problem, rather than supplying the data directly as Williams, Blowers and Goldberg (2004) illustrate in this example:

> **Before:** The following vapor pressure data is available: at 32° C, $P^*(pisa) = 0.08854$, at 34°C, $P^*(pisa) = 0.09603$. Estimate the heat of vaporization of water.

> **After:** Find vapor pressure data for H_2O and estimate the heat of vaporization of water. Give a full citation to the data you used and include a photocopy of the data itself. ("How to Incorporate IL" section)

Riley and Piccinino (2009) list assignments addressing information literacy skills as part of a Mass and Energy Balances course:

> a. Who developed the patent for the Clif shot litter leash? What earlier innovations did it build upon? (Requires a patent search.)

> b. Find two peer reviewed original research articles published in the last year about how climate change will affect hurricane frequency in the Atlantic. (Practices database search strategies and proper citation format).

c. What is ballast in a fluorescent light fixture? (Focuses on evaluation of sources; many definitions of ballast exist, but what sources are authoritative in this context?)

d. What is the National Science Foundation's budget for fiscal 2008? (Requires current knowledge that might be most appropriate for an Internet search, but conflicting reports require students to evaluate sources.)

e. Find a review of previous research on materials engineers' use of materials in space which are inspired by the natural world. (Practices database search strategies and citation, as well as distinguishing original research from review articles) ("Learning Strategies" section)

Single stand-alone library instruction sessions are challenging. Time constraints require that one be selective in which learning outcomes to address and recognize that students need to be proficient enough by the end of the session that they are able to construct a search statement, execute searches and evaluate the results at least at a beginner level. Starting the session with a quick overview of the process provides a context that is helpful to many students. A hand-out (see Appendix 2.2) helps provide the larger picture of the process and sets the stage for focusing on specific parts of the process: constructing a search strategy, executing the search in a database and importing the results into a bibliographic management program (such as RefWorks).

Another approach to the stand-alone session is a "research studio" in which students rotate through six stations and engage in exercises directly related to their assignment for a course. Each station focuses on particular skills such as search strategies, accessing a wide range of resources of information, and evaluation. The authors include details of the exercise, the related Standards performance outcomes for each station, and the worksheets for each of the stations. (Baratta, Chong, & Foster, 2011)

Worksheets, such as the one in Figure 2.1, assist students in refining a topic and creating a search strategy. Steps in the process the students work through include turning the topic into a question, identifying the major concepts involved, generating synonyms for each concept and, finally, constructing an appropriate search statement. Working through the steps enables the students to think more thoroughly about their topic and prepares them to narrow or broaden it as needed. Also, additional synonyms or alternate terms can be incorporated or substituted into the process as these terms and phrases are discovered by executing various iterations of the search statements.

Figure 2.1. Sample Search Statement Construction Worksheet		
1. State your topic as a question		
2. Separate into concepts		
3. Generate synonyms for each concept		
Concept #1	Concept #2	Concept #3
Synonym #1 for Concept #1	Synonym #1 for Concept #2	Synonym #1 for Concept #3
Synonym #2 for Concept #1	Synonym #2 for Concept #2	Synonym #2 for Concept #3
4. Sample Possible Search Statements: (Concept #1) AND (Concept #2) AND (Concept #3) (Concept #1 **OR** Syn #1 for Concept #1) **AND** (Concept #2 OR Synonym for Concept #2)		

Another model incorporates information literacy into a single course using several sessions. An example of this model is described by Parker and Godavari at the University of Manitoba where six classes of the Technical Communication Course are used for information literacy instruction. The sequence of the content in these classes is:

1. Overview
2. Developing a research question and citing references
3. Searching Google, Google Scholar, Google Books, and Google Alerts
4. Searching Engineering Village
5. Plagiarism and copyright
6. Additional engineering sources including patents, standards, technical reports and theses.

Some programs have incorporated information literacy into a program by sequentially building on the previous courses. At Villanova University, where five courses have explicit information literacy activities incorporated to accomplish 26 selected learning outcomes, the sophomore year term paper is a semester-long project with assignments submitted periodically throughout the semester. These assignments guide the students through the process beginning with topic selection and proceeding step-wise through searching, evaluation and summarizing, citing, outlining, and writing. Students also submit a written journal of their searching. The junior year activity is a case study, and the senior year includes evaluation of resources and a senior research paper (Welker, et al., 2010).

A program that has information literacy incorporated into the freshman, sophomore, junior and senior year is North Carolina State University as described by Nerz and Bullard (2006). The sequence is shown in Table 2.3.

Table 2.3. Information Literacy Integration at North Carolina State University	
Level	**Curricular Activities**
Freshman	• Library website • Database and catalog searching • Citing references
Sophomore	• Equation solving packages • Spreadsheets • Specific resources for Chemical Engineering with an assignment to research data for a specific compound and information on a technical topic
Junior	• Paper and presentation on a technical topic
Senior	• Construct a research map for the senior design project
Adapted from Nerz and Bullard (2006).	

Evaluation of information is essential and students need to develop habits of evaluating both the source and the information itself. Evaluation of sources involves applying the concepts of authority (what are the author's credentials and who is the publisher?), accuracy (is it facts or opinion?), bias, purpose, currency, coverage, and applicability to one's topic. Evaluation of data sources is especially important to engineers. McCullough (2006) lists a series of questions to guide students in developing critical thinking skills to evaluate potential information and data sources. These questions expand upon the evaluation of the source to include specifically addressing aspects of the information or data itself and are applicable to a wide variety of questions.

- What information sources are available for evaluation?
- Is the information directly applicable to the situation at hand?
- If not, how close is it to the current situation?
- What underlying assumptions have been made in the data?
- Is there any reason to suspect bias of any sort in this data source?
- How good is the evidence given by (or cited) in the source?
- Is there any potential conflict of interest?
- Is any significant data omitted?
- Are there any other data sources which should be consulted?
- Are there conflicting potential causes for the event?

(Incorporating Information Literacy into Existing Courses Section)

Numerous internet tools are available to connect with students at the point of need. Guides, tutorials, and "Ask a Librarian" chat links are ubiquitous on the internet. Many librarians have IM links on their research guides. In addition, links to library resources can be incorporated into course management software. Denick, Detweiler, Ray, Cebulski, and Bhatt (2009) described working with an interdisciplinary student group in a virtual environment using tools such as tutorials and a blog. The future will undoubtedly bring even more tools for easily working with students in a virtual environment.

STRATEGIES FOR SUCCESS IN ENGINEERING LIBRARIANSHIP
Professional Involvement

American Society for Engineering Education (ASEE) is a 12,000-member nonprofit professional organization whose mission is to enhance engineering and engineering technology education. Membership in the Engineering Libraries Division (ELD), one of the many divisions of ASEE, is very important to academic librarians. The ELDNET-l email listserv is a forum for discussion of relevant issues and an invaluable source of information and assistance to engineering librarians. ELD has a number of committees that provide opportunities to actively work with colleagues. The ELD division also sponsors workshops and sessions devoted to Information Literacy at the annual ASEE conference. These presentations and papers include examples of information literacy instruction content as well as rubrics for assessment (Engineering Libraries Division, 2011). Attendance at the ASEE Annual Conference also provides an opportunity for librarians to attend presentations relevant to information literacy by members of other divisions. Additional professional organizations for engineering librarians include the Special Libraries Association (SLA) Engineering Division, and the Science and Technology Section (STS) of the Association of College and Research Libraries (ACRL). STS publishes *Issues in Science and Technology Librarianship,* which includes articles and other materials on topics relevant to science and engineering/technology libraries.

Literature Sources

Two reference books for librarians that cover types of information sources as well as specific information sources for engineering are *Using the Engineering Literature* by Osif and *Information Sources in Engineering* by MacLeod and Corlett (see Additional Resources). These reference books include information about engineers and engineering and the information sources for the sub-disciplines. These compilations are good resources for becoming familiar with the vast content available. Librarians have developed excellent online research guides to resources for specific areas and topics in engineering. These research guides are excellent sources of compilations of selected resources on a wide variety of topics.

Standards

Engineering students frequently need industry or government standards. Standards are defined as "Common and repeated use of rules, conditions, guidelines or characteristics for products or related processes and production methods, and related management systems practices" (National Institute of Standards and Technology, 2012). They are created by hundreds of different organizations. The American National Standards Institute (ANSI) accredits U.S. organizations that develop standards and coordinates the creation, dissemination, and use of standards. ANSI is also the official United States representative to the International Standards Organization (ISO) and the International Electrotechnical Commission (IEC). The Institute of Electrical and Electronics Engineers (IEEE) and the American Society for Testing and Materials (ASTM) are two organizations accredited by ANSI that develop standards commonly needed by researchers and students.

Locating appropriate standards for a product or process can be time consuming. The web page, "How to Find Standards," developed by the National Institute of Standards (NIST) provides an excellent starting place for an overview of this process. It provides links to vendor online catalogs (e.g., NSSN, (the ANSI search engine for standards), IHS, TechStreet, SAI, and Document Center) that can be searched without charge. It also includes links to the government search engine for military standards, many of which are free full text online, as well as to databases of standards used by specific government agencies, such as Environmental Protection Agency (EPA) and the Food and Drug Administration (FDA) (NIST 2011). In addition to these resources, many individual organizations such as IEEE and ASTM have search engines available to identify the standards they developed. According to the NIST website, federal agencies adopt standards whenever possible, rather than create them in-house. NIST maintains the Standards.gov website to provide information specifically about standards used in government. The website has extensive information about standards, including types of standards and resources for finding standards. It also provides a section on regulations and the use of agency-developed or private sector-developed standards in government regulations. The "Standards Incorporated by Reference" (SIBR), another section of the Standards.gov website, is a database of the standards referenced in the Code of Federal Regulations (CFR). Research guides created by librarians are useful tools to assist in locally identifying and obtaining standards (see Additional Resources).

Technical Reports

Technical Reports, an important source of information for engineering, are the result of research funded by the government or private entities. In addition to the government, technical reports are published by universities or industry. Science.gov is a starting point for access to 50 databases and over 2100 selected websites in 18 scientific and technical organizations from 14 federal agencies (Science.gov 2011). Included in this list are Department of Energy (DOE), National Technical Information Service (NTIS) and the Environment Protection Agency (EPA). One can also search the website of the specific agency for technical reports. The Technical Report Archive and Image Library (TRAIL), a collaborative project that digitizes older technical reports issued by government agencies, is a good source for technical reports issued prior to 1975.

Other Sources

Other areas important to engineers are business information and patents (addressed in chapter 6 of this book). Engineering students may be asked to research an industry and/or a company. Arizona State University (ASU) has developed a flow chart for obtaining specific company and industry information (LaFaro, 2010). Biomedical engineering students at ASU find this flowchart arrangement and content very useful in researching company and industry information (personal observation).

CONCLUSION

The opportunities for teaching information literacy grow with the increasing awareness of the importance of information literacy in engineering. The ability to know when one needs information, as well as how to obtain, evaluate, and use the information are skills critical to engineers. These skills are best taught sequentially and incorporated into the curriculum making use of learning strategies that engage students and provide them with a purpose. Many resources exist to assist new engineering librarians in developing learning outcomes, assessment tools, and teaching strategies. This chapter has presented only a few of the resources available. The literature of engineering librarianship contains many more examples describing the process of incorporating information literacy at the course level and the program level. Virtual tools expand the contact possibilities that librarians have with students and the variety and ease of use of these tools facilitate more point-of-need assistance than ever before. Working with engineering students and faculty, one can appreciate that information literacy and the critical thinking it requires contribute significantly to the education of future engineers.

REFERENCES

ABET. (2011). Accreditation criteria, policies, and procedures. Retrieved from http://abet.org/accreditation-criteria-policies-documents/

ABET. (2011, July 26) Self-study questionnaire templates, 2012–2013 review cycle. Retrieved from http://abet.org/download-self-study-templates/

ACRL. (2000). Information literacy competency standards for higher education. Retrieved from http://www.ala.org/acrl/sites/ala.org.acrl/files/content/standards/standards.pdf.

ALA/ACRL/STS Task Force on Information Literacy for Science and Technology [STS-TFILST]. (2006). Information literacy standards for science and engineering/technology. Retrieved from http://www.ala.org/acrl/standards/infolitscitech

American Society for Engineering Education (ASEE). (2011). Homepage. Retrieved from http://www.asee.org/

Angelo, T. A., & Cross, P. (1993). Classroom assessment techniques: A handbook for college teachers (2nd ed.). San Francisco, CA: Jossey-Bass.

Baratta, M., Chong, A., & Foster, J. A. (2011, June). The research studio: Integrating information literacy into a first year engineering science course. Paper presented at the 2011 ASEE Annual Conference and Exposition, Vancouver, BC, Canada. Retrieved from http://www.asee.org/search/proceedings

Briedis, D. M. (2010). Assessment: Developing rubrics [Webinar]. Retrieved from http://www.abet.org/developing-rubrics/

College of Engineering and Architecture and the Center for Teaching, Learning and Technology, Washington State University. (2009, April 28). Guide to rating ABET professional skills. Retrieved from https://assessment.wsu.edu/programs/CEA

Denick, D., Bhatt, J., & Layton, B. (2010 June). Citation analysis of engineering design reports for information literacy assessment. Paper presented at the 2010 ASEE Annual Conference and Exposition, Louisville, KY. Retrieved from http://www.asee.org/search/proceedings

Denick, D., Detweiler, J., Ray, C., Cebulski, A., & Bhatt, J. (2009, June). Library-smart house collaboration for information literacy development. Paper presented at the 2009 ASEE Annual Conference and Exposition, Austin TX. Retrieved from http://www.asee.org/search/proceedings

Engineering libraries division (ELD). ASEE. (2011). Retrieved from http://depts.washington.edu/englib/eld/

Felder, R. M. (2010 September 27). "Are Learning Styles Invalid? (Hint: No!)." On-Course Newsletter. Retrieved from http://www4.ncsu.edu/unity/lockers/users/f/felder/public/Papers/LS_Validity(On-Course).pdf

Felder, R. M., & Silverman, L. K. (1988). Learning and teaching styles in engineering education. Engineering Education, 78, 674–681. Retrieved from http://www4.ncsu.edu/unity/lockers/users/f/felder/public/Papers/LS-1988.pdf

Formative vs. summative assessment. (n.d.) Retrieved from http://www.cmu.edu/teaching/assessment/howto/basics/formative-summative.html

Ira A. Fulton Schools of Engineering, Arizona State University, Fulton Undergraduate Research Initiative (FURI). (n.d.). Undergraduate research. Retrieved from http://engineering.asu.edu/furi

LaFaro, L. (2010). Research process flowcharts. Retrieved from http://libguides.asu.edu/content.php?pid=17228&sid=116716

MacAlpine, B., & Uddin, M. (2009, June). Integrating information literacy across the engineering design curriculum. Paper presented at the 2009 ASEE Annual Conference and Exposition, Austin TX. Retrieved from http://www.asee.org/search/proceedings

Ira A. Fulton Schools of Engineering, Arizona State University, Motivated Engineering Transfer Scholars (METS). (n.d.) Homepage. Retrieved from http://mets.engineering.asu.edu/

McCullough, C. L. (2006). *Information literacy: A critical component in engineering practice in the twenty-first century*. Paper presented at the 2006 *ASEE Southeast Section Meeting*, Tuscaloosa, AL. Retrieved from http://155.225.14.146/asee-se/proceedings/ASEE2006/P2006049MCC.pdf

National Institute of Standards and Technology. (2012). Standards.gov. Retrieved from http://standards.gov/

Nelson, M. S., & Fosmire, M. (2010, June). *Engineering librarian participation in technology curricular redesign: Lifelong learning, information literacy, and ABET criterion 3*. Paper presented at the 2010 ASEE Annual Conference and Exposition, Louisville, KY. Retrieved from http://www.asee.org/search/proceedings

Nerz, H., & Bullard, L. (2006, June). *The literate engineer: Infusing information literacy skills throughout and engineering curriculum*. Paper presented at the 2006 ASEE Annual Conference and Exposition, Chicago, IL. Retrieved from http://www.asee.org/search/proceedings

Nerz, H., & McCord, M. G. (2008). Information literacy rubric. Retrieved from http://www.lib.ncsu.edu/instructiontoolkit/document.php?doc=82

NIST. (2011). How to find standards. Retrieved from http://gsi.nist.gov/global/index.cfm/L1-5/L2-44/A-171

Parker, A. & Godavari, S. N. (2011, June). *The value of direct engagement in a classroom and a faculty: Using the liaison librarian model to integrate information literacy into a faculty of engineering*. Paper presented at the 2011 ASEE Annual Conference and Exposition, Vancouver, BC, Canada. Retrieved from http://www.asee.org/search/proceedings

Riley, D., & Piccinino, R. (2009, June). *Integrating information literacy into a first year mass and energy balances course*. Paper presented at the 2009 ASEE Annual Conference and Exposition, Austin TX. Retrieved from http://www.asee.org/search/proceedings

Riley, D., Piccinino, R., Moriarty, M., & Jones, L. (2009, June). Assessing information literacy in engineering: Integrating a college-wide program with ABET-driven assessment. Paper presented at the 2009 ASEE Annual Conference and Exposition, Austin TX. Retrieved from http://www.asee.org/search/proceedings

Rogers, G. (2010). *Assessment: Choosing assessment methods: Webinar and Slides*. Retrieved November 25, 2011, from http://www.abet.org/choosing-assessment-methods/

Science.gov. (2011). Homepage. Retrieved from http://www.science.gov/index.html

Shuman, L. J., Besterfield-Sacre, M., & McGourty, J. (2005). The ABET "professional skills" can they be taught? Can they be assessed? *Journal of Engineering Education, 94*, 41–55.

TRAIL: Technical report archive & image library. Retrieved November 25, 2011, from http://www.crl.edu/grn/trail

Welker, A., Fry, A., McCarthy, Li, & Komlos, J. (2010). An integrated approach to information literacy instruction in civil engineering. *Mid Atlantic ASEE Conference*, Villanova University. Retrieved from http://jee.asee.org/documents/sections/middle-atlantic/fall-2010/01-An-Integrated-Approach-to-Information-Literacy-Instruction-i.pdf

Wertz, R. E. H., Ross, M. C., Purzer, S., Fosmire, M., & Cardella, M. E. (2011, June). *Assessing engineering students' information literacy skills: An alpha version of a multiple -choice instrument*. Paper presented at the 2011 ASEE Annual Conference and Exposition, Vancouver, BC, Canada. Retrieved from http://www.asee.org/search/proceedings

Wiggins, G. P., & McTighe, J. (2005). *Understanding by design* (Exp. 2nd ed.). Alexandria, VA: Association for Supervision and Curriculum Development.

Williams, B., Blowers, P., & Goldberg, J. (2004, June). *Integrating information literacy skills into engineering courses to produce lifelong learners.* Paper presented at the 2004 American Society for Engineering Education Annual Conference and Exposition, Salt Lake City, UT. Retrieved from http://www.asee.org/search/proceedings

ADDITIONAL RESOURCES

Osif, B. A. (Ed.). (2012). *Using the engineering literature* (2nd ed.) Boca Raton, FL: CRC Press.

MacLeod, R. A., & Corlett, J. (2005). *Information sources in engineering.* München: K. G. Saur.

Shackle, L. (2011). *Standards (engineering): A guide to locating technical standards and a list of standards available in Noble Library.* Available at http://libguides.asu.edu/content.php?pid=8556&sid=54945

Standards incorporated by reference (SIBR) database. (2009). Available at http://standards.gov/sibr/query/index.cfm

APPENDIX 2.1. SAMPLE "CATS" USED AS FORMATIVE FEEDBACK

These are sample "CATS" questions that can be used to gather information to improve teaching after a single session instruction. Because of time constraints, using 3–4 selected questions is optimal with a final CAT being space for "Additional Comments."

1. What did you learn that you didn't already know?
2. What was the clearest?
3. What was unclear?
4. If you don't find any—(patents, articles, or other depending on the topic of the instruction)—on your topic, what might you do next?
5. Of what you learned today, what will be the most helpful?
6. Before today, what databases, resources, or methods have you used to find articles from journals or conferences?
7. What would you like to have covered that was not covered?

Note: always include a space for "Additional Comments"
Sample Feedback Form:

Feedback : Course number
Semester
Year

1. Of what you learned today, what will be the most helpful?

2. What was the clearest?

3. What was unclear?

4. What would you like to have covered that was not covered?

5. Additional comments:

Thanks!

APPENDIX 2.2. LITERATURE REVIEW OF A TOPIC

Secondary Literature: Books and Specialized Encyclopedias
Review Articles

Primary Literature   Original Research –Journal and Conference
Literature Databases Patents

1.   State the topic as a question

2.   Who would create this information? Or do I have to create it myself

3.   Where would it be located?
Examples: books, published in journals or conference proceedings? proprietary? patents? Government publications? Technical reports? Web sites? Personal communication? Other?

4.   Select resources

5.   How do I retrieve it? If using databases, how does the search engine work? Cited reference searching.

6.   Construct Search Statement(s) and execute searches
Multiple searches: broader, narrower, and synonym keyword searches "Controlled Vocabulary" searches may be more specific, but may miss newer concepts.

7.   Evaluate results using criteria

8.   How do I organize my searches and results?
Record resources
Record search statements
Import results into folders in "RefWorks" or other bibliographic management tools such as "Bibtex"
Manually enter other references into RefWorks or other bibliographic management software

9.   How do I keep up to date?
table of contents alerts
Email alerts
RSS feeds

IMPORTANT INFORMATION
LITERACY STANDARDS FOR
LIFE SCIENCES

1. The information literate student determines the nature and extent of the information needed.
    3. Has a working knowledge of the literature of the field and how it is produced.
        a. Knows how scientific, technical, and related information is formally and informally produced, organized, and disseminated.
        b. Recognizes that primary, secondary, and tertiary sources vary in importance and use with each discipline.

2. The information literate student acquires needed information effectively and efficiently.
    1. Selects the most appropriate investigative methods or information retrieval systems for accessing the needed information.
        b. Investigates the scope, content, and organization of information retrieval systems.
    2. Constructs and implements effectively designed search strategies.
        f. Follows citations and cited references to identify additional, pertinent articles.

3. The information literate student critically evaluates the procured information and its sources, and as a result, decides whether or not to modify the initial query and/or seek additional sources and whether to develop a new research process.
    2. Selects information by articulating and applying criteria for evaluating both the information and its sources.
        a. Distinguishes between primary, secondary, and tertiary sources, and recognizes how location of the information source in the cycle of scientific information relates to the credibility of the information.

5. The information literate student understands that information literacy is an ongoing process and an important component of lifelong learning and recognizes the need to keep current regarding new developments in his or her field.
    2. Uses a variety of methods and emerging technologies for keeping current in the field.

IMPORTANT INFORMATION LITERACY STANDARDS FOR HEALTH SCIENCES

2. The information literate student acquires needed information effectively and efficiently.
   2. Constructs and implements effectively designed search strategies.
      b. Identifies keywords, synonyms and related terms for the information needed and selects an appropriate controlled vocabulary specific to the discipline or information retrieval system.
      d. Constructs a search strategy using appropriate commands for the information retrieval system selected (e.g., Boolean operators, truncation, and proximity for search engines; internal organizers such as indexes for books).

LIFE AND HEALTH SCIENCES

CHAPTER 3

Elizabeth (Betsy) Hopkins
Life Sciences Librarian
Brigham Young University

Information literacy in the life and health sciences is a dynamic field, with challenges, opportunities, and rewards for the successful practitioner. This chapter will describe the big picture of information literacy in these disciplines, list relevant performance indicators from the ALA/ACRL/STS Task Force on Information Literacy for Science and Technology [STS-TFILST] (2006) Information Literacy Standards for Science and Technology (hereafter Standards), and provide some practical advice for life and health sciences librarians and librarians with instructional responsibilities in those disciplines. The focus is on undergraduates at research universities, although many principles and strategies will apply to other constituent groups and types of academic institutions.

THE LANDSCAPE: CHARACTERISTICS, CHALLENGES, AND OPPORTUNITIES

Encompassing a broad range of study, the life sciences examine life in all its forms and at all levels, from microscopic protein pathways functioning in a cell to whole organisms and their interactions with other organisms and their environment. It includes a number of specific sub-disciplines, from molecular biology and physiology to zoology and botany. Researchers work in the lab or in the field, and often both.

As in any field, student characteristics vary somewhat. There is a large group of life sciences majors who are pre-professional; these students will go on to further education in biomedical professional programs (e.g., medical, dental, pharmacy, optometry). Another group of life sciences majors will complete research-based programs in graduate school, obtaining MS and PhD degrees. Pre-professional students, particularly those headed for medical school, tend to be competitive and high-achieving. Many students, regardless of track, will want to gain laboratory experience by working in labs run by faculty researchers.

Professors in the life sciences tend to be independent, believing that they don't need assistance from the library. In fact, many faculty in these disciplines view the library as a content provider only, and it can be difficult to convince them of the value of information literacy for themselves and their students.

In the life sciences, peer-reviewed journals are the most important form of information dissemination. Monographs are used rarely, if at all. Research assignments typically require the use of peer-reviewed literature. Lower-division courses that meet the institution's science general education requirements have tended towards relaxing the standards for using high-level information sources. Some require students to use the Internet and focus assignments on evaluating sources in an attempt to parallel real-life learning. Major-based, upper-division courses require peer-reviewed, mostly primary sources. Primary sources report on experiments conducted by the authors, as opposed to summaries or other secondhand reports.

The most important databases in the life sciences are PubMed (Medline), Biosis Previews, Zoological Record and Web of Science (Scopus is used by some institutions in place of or along with Web of Science). Smaller, discipline-specific databases, such as Agricola, Environmental Science, GREENR, and SPORTDiscus, are useful for specialty areas like agriculture, environmental science and exercise science. Pubmed (www.pubmed.gov) has an excellent search algorithm and is provided free to all by the National Library of Medicine. Pubmed's highly effective keyword search (that automatically maps to Medical Subject Headings, or MeSH terms) allows for quick retrieval of relevant results. Subject headings in most life sciences databases are not as important as in some other disciplines, because, quite often, they are too broad to be useful for typical searches. Taxonomic categories in BIOSIS Previews and Zoological Record are the only exception, as they can help narrow search results to a specific species.

The health sciences are related to the life sciences and often overlap with them. The health sciences focus on the applied life sciences and how to ensure human health. They may include nursing, nutrition, dietetics, public health, health promotion and wellness, athletic or fitness training, and pre-physical and pre-occupational therapy. Health sciences students tend to be less competitive or high-achieving than some in the life sciences, and professors in these fields tend to be more open to assistance from the library. Nursing in particular has a high level of engagement with the library, as demonstrated by a wide body of literature (see Flood et al., 2010; Guillot et al., 2010; Phillips & Bonsteel, 2010). A wider variety of sources and databases are available in the health sciences, including such smaller, specialized resources as AltHealthWatch and PEDro. Clinical sources such as UpToDate are more important, as are evidence-based sources, including the Cochrane Library and National Guidelines Clearinghouse (http://www.guideline.gov/).

Both challenges and opportunities in information literacy abound in the life sciences. Faculty and student resistance is often a problem. Because professors believe their students already have information literacy skills and feel that their classes are already full to the brim with content, they don't want to make room for information literacy instruction. In addition, students often believe they already know how to do literature research. Another challenge is structural. The presence of information literacy instruction in a given course is often based on the individual relationships librarians build with professors, instead of structural integration into the life sciences curriculum. Personnel turnover of faculty and librarians, as well as arbitrary changes in research assignments, can make information literacy integration transient. McGuinness (2007) reviews the literature related to integrating information literacy into the curriculum and concludes that partnering with an academic champion (one academic who is supportive of information literacy) does not lead to long-term integration. She advocates for top-down planning and targeting institution-wide initiatives, which are more likely to result in long-term information literacy programs. Yet another challenge is the high volume of students in the life sciences. Scores of students are attracted to biology, and high enrollment can make instruction difficult for library personnel, who are often limited in number.

Despite these challenges, there are a number of opportunities for information literacy in the life sciences. Professors in the life sciences give research assignments as one method to help students develop communication skills. In addition, in recent years, the evaluation of information has increased in importance. Faculty are expecting their students to find quality information and understand why it is valuable. This takes the focus off the logistics and navigation of a library website. Similarly, it has become more important that students understand the information landscape of their chosen fields. This allows them to navigate to the best sources for their information need. Recently, Ferrer-Vinent and Carello (2011) explored another opportunity when they demonstrated that life sciences information literacy instruction led to skill retention three years after the instruction. This research supports the idea that life sciences librarians can make a long-term difference. Taken together, these opportunities provide an opening for life sciences librarians to integrate information literacy into the curriculum.

IMPORTANT STANDARDS

The Standards provide a framework for all science librarians, although some performance indicators are more relevant to the life and health sciences than others. This section presents the indicators most important for students in these fields. The Standards are written in a language librarians understand, so it is important to translate the standards and performance indicators into the more general language of research skills, because this is what faculty understand. The following discussion reflects some useful translations.

This selection of standards is shaped by a teaching philosophy of restricting instruction to the most important content, the "less is more" concept, as librarians often have limited face time with students. One-shot sessions are common, but one session doesn't provide enough time to effectively teach students all the skills librarians think are valuable. Librarians feel tempted to cram sessions full of everything they know that would help students; however, they must resist the urge to lecture for 50 minutes and "fit everything in." Students become alienated and do not learn. Instead of trying to present everything, start with the assumption that all students can retrieve results from databases, and then focus on the skills that will help them efficiently sift and sort those results to find sources they can use for their assignments. Try to build in time for hands-on searching and active learning exercises. Based on this philosophy, the following list of standards is not comprehensive, but instead reflects those that are most important to the life and health sciences.

Life Sciences

Performance indicator 1.3, and in particular 1.3.a, refers to students' knowledge of the life sciences literature. Students need to understand the flow of information in the discipline, from laboratory experiment or observation of the natural world to primary journal article, then on to review articles and other sources. Performance indicators 1.3.b and 3.2.a both relate to primary, secondary, and tertiary sources. Taken together, the translation of these indicators is that students should be able to identify sources by type and recognize the advantages of each.

Understanding the flow of information in the life sciences and characteristics of primary, secondary, and tertiary sources is an important skill for life sciences students, because each type of source matches different information needs. Often students need to start a research project by looking for relevant review articles, because it helps them understand increasingly technical and specific primary literature. Primary literature reports new information, and includes details about methodology and data, but review articles are critical for summarizing trends and bringing together current knowledge in a given field. Tertiary sources are excellent for topic development and general background information. Understanding these distinctions will allow students to navigate directly to the types of sources that align with their current needs.

Relevant in all disciplines, performance indicator 2.1.b recommends that students understand the focus of each database and be able to select those appropriate for their information needs. Student researchers must select appropriate databases, and instruction about the subject coverage and selection criteria of each available source will provide the necessary information for this choice. Becoming familiar with included fields, limits, and other aspects of database organization will allow students to use advanced search interfaces effectively.

Performance indicator 2.2.f notes the significance of bibliographic searching, or cited reference searching. Students use this skill effectively when they examine reference lists of other suitable sources. Many researchers do this by looking at bibliographies in relevant articles; however, Web of Science, Scopus and Google Scholar allow users to view cited references for most articles. The dynamic linking of titles and abstracts is a boon to students as they can more easily identify relevant articles using abstracts. This standard is particularly meaningful in biology because it provides another way to find relevant articles that keyword searching may have missed. Although not explicitly stated, this indicator also encompasses the importance of using citation numbers to identify high impact papers. For example, in Web of Science, it is possible to sort a results set by the number of times other articles have cited it. This allows students to identify those landmark articles that must not be missed by anyone investigating the subject.

Performance indicator 5.2 is important for those life sciences students in laboratory groups or on long-term projects, because it notes the value of currency in the life sciences. The search alert is an excellent tool for keeping up to date. Life sciences students have a particular need for current awareness services because lab research is by its nature outside of the typical semester framework. Students in a lab may work on a project for a year or two, and they will need to stay abreast of relevant research.

Health Sciences

In the health sciences, there are two additional outcomes of particular consequence, 2.2.b and 2.2.d. The first notes that students need to identify and use subject headings relevant to their topics. This standard is important for the health sciences, because a wider variety of databases are available in this area, and many require using subject heading searches in order to produce effective and comprehensive results. For example, using subject headings in CINAHL generally leads to more relevant results than keyword searching. The second indicator, 2.2.d, notes the importance of applying Boolean operators and truncation correctly. Many specialized health sciences databases require careful search construction for relevant results. In addition, health sciences students are typically less familiar with these concepts.

Some disciplines have their own standards, created by professional associations or other relevant organizations. Nursing, for example, has several guidelines and recommendations for information literacy in the field. These include the American Association of Colleges of Nursing's (AACN) *Essentials of Baccalaureate Nursing Education* (2008), the Institute of Medicine's report on the future of nursing (2010), the National League for Nursing's position statement on informatics (2009), and the Technology Informatics Guiding Education Reform Initiative (2007). Each speaks to the importance of information literacy for nurses and nursing education.

PRACTICAL ADVICE

In any field, tips from other practicing professionals are helpful. This section provides practical advice on integrating information literacy into the life sciences curriculum, incorporating active learning, and working with faculty in the life sciences. Lastly, the section includes some advice for librarians without a science background.

Integrating Information Literacy

A variety of approaches is useful when working to integrate information literacy within a curriculum. An effective way for librarians and faculty to think about these options is as a menu, similar to what one might find in a restaurant. This approach has been used by a number of institutions; for a summary and analysis of instructional menus, see Benjes-Small, Dorner, and Schroeder (2009). Consider developing a list of options that may be used on their own or in combination with other approaches. Circumstances of the course, the professor, and student needs will dictate the appropriate combination of strategies. These run the gamut, from the simple, and perhaps less effective, mention of available librarian help in the syllabus to several in-class library presentations. Options for varying levels of involvement include:

- Mention librarian/reference and information desks/open labs in syllabus
- In-class instruction, from 5 minutes to several hours in length
- Out of class instruction: required, extra credit, or optional
- Instruction in laboratory or recitation sections
- Several instruction sessions (see Petzold et. al 2010)
- Online course guides
- Tutorials (see Schroeder 2010)
- Blackboard or other course management systems (see DaCosta and Jones 2007)
- Assignments (see Ferrer-Vinent and Carello 2008)
- Drop-in research labs: with or without associated course credit

Several examples used in practice are presented in Table 3.1.

Incorporating Active Learning

Instruction of any kind is more effective when it is active rather than passive. Table 3.2 lists some techniques that may be used when teaching science information literacy. Most of these strategies can be used to teach a number of principles, although the table lists examples for standards previously identified as important in the life sciences. The STS Information Literacy wiki (STS Information Literacy Committee (STS-IL), 2011) also has a collection of active learning tips based on the performance indicators for each of the Standards.

Table 3.1. Examples of Integrating Information Literacy into the Curriculum	
Type of Course	Strategies
Freshman-level Introductory Biology, with variety of instructors and assignments	Tailored to each professor, including: • 20-minute in class presentation on finding peer reviewed literature for environmental solution paper • 2-hour in class presentation on evaluating sources and library basics for brief, weekly, subject-based writing assignments • 1-hour out of class, optional presentations on flow of information and finding peer reviewed literature for traditional research paper assignment; extra credit received for attending • Section-specific online course guides
300-level Tissue Biology, with research paper assignment	Required 50-minute library instruction session during the first week of course lab session. In library classrooms.
Nursing Curriculum (see Hopkins et. al 2011)	2nd semester Gerontology class: 10 minute introduction to the nursing librarian and to CINAHL as the main nursing database 3rd semester Nursing Research class: • First 2-hour session: 40-minute lecture, the remainder hands-on work time • Second 2-hour session: case study, with mixed hands-on work time, class discussion, and chunked lecture 4th semester Nursing Ethics class: 2-hour work session in the library, no lecture

Working with Life Sciences Faculty

Opportunities to teach in life sciences courses are more likely to come from systematically built relationships with the life sciences faculty. Try these strategies as good starting points:

- Make appointments with all faculty in assigned departments and schools. Start with the department chair and other administrators. Ask about research assignments and related learning outcomes. Offer tailored suggestions for how you can help students develop research skills.
- Visit or regularly attend department faculty meetings. Faculty meetings can be a good place for department-wide curriculum discussions.
- Pay attention to the reference questions you receive and ask students about their assignments and professors. Then follow up with those faculty, discuss student needs, and offer assistance.
- Standardize across a department or college curriculum if possible. Without standardization, students may be exposed to multiple settings for the same library lecture, which they find annoying and inefficient.

Table 3.2: Active Learning Strategies

Performance Indicator	Active Learning Tip	Example
1.3 and 3.2.a	Questioning. This simple technique breaks up a lecture while still delivering information. Be sure to wait for responses from your students.	"What are the advantages of a secondary source as compared to a primary source?"
1.3 and 3.2.a	Pop quiz.	Display abstracts and require students to identify the source as primary, secondary, or tertiary.
2.1.b	Case studies (particularly effective in the health sciences).	Present a scenario and ask students to investigate which database is best suited to the information need.
2.2.b	Worksheets. Think about a process you want the students to follow and create a worksheet to guide them through the process.	Identifying and searching with subject headings. Worksheet might include space for initial keywords, number of associated results, a list of subject headings, and the number of results from this new search.
2.2.f	An assignment. This works best if you can tie it to course credit. Build work time into your session.	Ask students to identify high impact articles using times cited information.

- Obtain copies of course syllabi and research assignments where possible. Some departments keep a central copy of all syllabi.
- Ask to be added to faculty email distribution lists for the college and department. This will keep you informed about initiatives and trends among the faculty.
- Become friendly with department secretaries and other support staff. These key personnel can also help you stay informed.

For the Librarian without a Science Background

Science librarians without a science background can use several techniques to provide effective services to their assigned departments. First, realize that many upper-level students will be able to explain the science related to their questions. Even librarians with science backgrounds cannot be familiar with everything and must rely on students to explain the basics of their information need. Second, identify regular assignments used by faculty at your institution. Improve your knowledge of these topics by reading several review articles and/or a popular book, if available. Third, stay aware of major developments in science by watching the major publications in the field. The life sciences are splintered into many sub-disciplines, but news articles in the journals *Science* and *Nature* can help. Take advantage of table of contents alerts for more specific fields. Fourth, observe trends within science librarianship. *Science and Technology Libraries* and *Issues in Science and Technology Libraries* are good publications to monitor. In the health sciences, try the *Journal of the Medical Library Association.*

CONCLUSION

Information literacy in the life sciences is an engaging discipline, with numerous opportunities. Despite some challenges from life sciences students and faculty, an increased emphasis on understanding the flow of information in the field has provided an opening for librarians to have a significant impact on student learning. Relevant performance indicators and outcomes from the Standards for the life sciences highlight skills for evaluating and managing results sets. Specifically, students should be able to match the type of source, primary or secondary, to their particular information need. In addition, students should be able to use cited reference searching to identify high impact papers. Indicators important for the health sciences include the use of Boolean operators, truncation, and subject headings. Life sciences librarians both new and experienced will benefit from practical advice on incorporating information literacy into the curriculum, integrating active learning techniques, and working with faculty. Lastly, a few suggestions for librarians without a science background can help them be successful. Creative and effective life sciences librarians will equip students with skills that will benefit them throughout their careers as scientists and biomedical professionals.

REFERENCES

ALA/ACRL/STS Task Force on Information Literacy for Science and Technology [STS-TFILST]. (2006). Information literacy standards for science and engineering/technology. Retrieved from http://www.ala.org/acrl/standards/infolitscitech

American Association of Colleges of Nursing. (2008). *The essentials of baccalaureate nursing education for professional nursing practice.* Washington, DC: American Association of Colleges of Nursing.

Benjes-Small, C., Dorner, J. L., & Schroeder, R. (2009). Surveying libraries to identify best practices for a menu approach for library instruction requests. *Communications in Information Literacy, 3,* 31–44. Retrieved from http://www.comminfolit.org/index.php?journal=cil&page=article&op=view&path[]=Vol3-2009AR2

DaCosta, J. W., & Jones, B. (2007). Developing students' information and research skills via blackboard. *Communications in Information Literacy, 1,* 16–25. Retrieved from http://www.comminfolit.org/index.php?journal=cil&page=article&op=view&path[]=Spring2007AR2

Ferrer-Vinent, I., & Carello, C. A. (2008). Embedded library instruction in a first-year biology laboratory course. *Science & Technology Libraries, 28,* 325–351. doi:10.1080/01942620802202352

Ferrer-Vinent, I., & Carello, C. A. (2011). The lasting value of an embedded, first-year, biology library instruction program. *Science & Technology Libraries, 30,* 254–266. doi:10.1080/019 4262X.2011.592789

Flood, L. S., Gasiewicz, N., & Delpier, T. (2010). Integrating information literacy across a BSN curriculum. *Journal of Nursing Education, 49,* 101–104. doi:10.3928/01484834-20091023-01

Guillot, L., Stahr, B., & Meeker, B. J. (2010). Nursing faculty collaborate with embedded librarians to serve online graduate students in a consortium setting. *Journal of Library & Information Services in Distance Learning, 4,* 53–62. doi:10.1080/15332901003666951

Hopkins, B., Callister, L.C., Mandleco, B., Lassetter, J., & Astill, M. (2011). Librarians as partners of the faculty in teaching scholarly inquiry in nursing to undergraduates at Brigham Young University. *Science and Technology Libraries, 30,* 267–276. doi:10.1080/019426 2X.2011.593416

Institute of Medicine. (2010). *The future of nursing: Leading change, advancing health.* Retrieved from http://www.iom.edu/Reports/2010/The-Future-of-Nursing-Leading-Change-Advancing-Health.aspx

McGuinness, C. (2007). Exploring strategies for integrated information literacy. *Communications in Information Literacy, 1,* 26–38. Retrieved from http://www.comminfolit.org/index.php?journal=cil&page=article&op=view&path[]=Spring2007AR3

National League for Nursing. (2009). *Position statement: Preparing the next generation of nurses to practice in a technology-rich environment: An informatics agenda.* Retrieved from http://www.nln.org/aboutnln/PositionStatements/informatics_052808.pdf

Petzold, J., Winterman, B., & Montooth, K. (2010). Science seeker: A new model for teaching information literacy to entry-level biology undergraduates. *Issues in Science & Technology Librarianship,* (63), 49–58. Retrieved from http://www.istl.org/10-fall/refereed2.html

Phillips, R. M., & Bonsteel, S. H. (2010). The faculty and information specialist partnership: Stimulating student interest and experiential learning. *Nurse Educator, 35,* 136–138. doi:10.1097/NNE.0b013e3181d95090

Schroeder, H. (2010). Creating library tutorials for nursing students. *Medical Reference Services Quarterly, 29,* 109–120. doi:10.1080/02763861003723135

STS Information Literacy Committee (STS-IL). (2011). Science information literacy [wiki]. Retrieved from http://wikis.ala.org/acrl/index.php/Science_Information_Literacy

Technology Informatics Guiding Education Reform Initiative. (2007). *The TIGER initiative: Evidence and informatics transforming nursing: Three year action steps toward a 10-year vision.* Retrieved from http://tigersummit.com/

IMPORTANT INFORMATION LITERACY STANDARDS FOR CHEMISTRY

1. The information literate student determines the nature and extent of the information needed.
   2. Identifies a variety of types and formats of potential sources for information.
   3. Has a working knowledge of the literature of the field and how it is produced.
   4. Considers the costs and benefits of acquiring the needed information.

2. The information literate student acquires needed information effectively and efficiently.
   1. Selects the most appropriate investigative methods or information retrieval systems for accessing the needed information.
   2. Constructs and implements effectively designed search strategies.
   3. Retrieves information using a variety of methods.
   4. Refines the search strategy if necessary.
   5. Extracts, records, transfers, and manages the information and its sources.

3. The information literate student critically evaluates the procured information and its sources, and as a result, decides whether or not to modify the initial query and/or seek additional sources and whether to develop a new research process.
   1. Summarizes the main ideas to be extracted from the information gathered.
   2. Selects information by articulating and applying criteria for evaluating both the information and its sources.
   3. Synthesizes main ideas to construct new concepts.
   4. Compares new knowledge with prior knowledge to determine the value added, contradictions, or other unique characteristics of the information.
   5. Validates understanding and interpretation of the information through discourse with other individuals, small groups or teams, subject-area experts, and/or practitioners.

4. The information literate student understands the economic, ethical, legal, and social issues surrounding the use of information and its technologies and either as an individual or as a member of a group, uses information effectively, ethically, and legally to accomplish a specific purpose.

IMPORTANT INFORMATION LITERACY STANDARDS FOR CHEMISTRY

1. Understands many of the ethical, legal and socio-economic issues surrounding information and information technology.
2. Follows laws, regulations, institutional policies, and etiquette related to the access and use of information resources.
3. Acknowledges the use of information sources in communicating the product or performance.
4. Applies creativity in use of the information for a particular product or performance.
5. Evaluates the final product or performance and revises the development process used as necessary.
6. Communicates the product or performance effectively to others.

5. The information literate student understands that information literacy is an ongoing process and an important component of lifelong learning and recognizes the need to keep current regarding new developments in his or her field.

CHEMISTRY

Olivia Bautista Sparks
Chemistry Librarian
Arizona State University

INTRODUCTION

Chemistry is a central science and is the foundation for many technologies, including nanotechnology and genetically modified organisms. Popular culture has depicted chemistry as bubbling flasks and liquids changing color in test tubes; however, chemistry is not always confined to the lab bench. It plays a role in the rust and lime buildup that can accumulate on an old car after exposure to the elements. Chemistry is involved in powering the human body through a series of chemical reactions that begin once food enters the mouth. Chemistry is also a challenging subject that is used as a measure of professional potential in medically related fields. Many admissions exams (e.g., Medical College Admissions Test (MCAT), Optometry Admission Test (OAT), Pharmacy College Admission Test (PCAT)) assess the student's proficiency in chemistry, and the scores received on these exams are an important consideration for admission into Medical, Optometry or Pharmacy School. With this added pressure, students often approach chemistry courses with anxiety and trepidation.

CHEMICAL INFORMATION INSTRUCTION AND THE STANDARDS

There is a long history of Chemical Information for-credit courses taught by librarians, faculty or a combination of both (Garritano & Culp, 2010). In reality, these types of courses are not found on every campus and the main challenge facing librarians is where and how to insert information literacy into the chemistry curriculum.

The American Chemical Society (ACS) understands the importance of chemical information resources and chemical information literacy; chemistry programs accredited by the ACS Committee on Professional Training (CPT) must have access to chemical information resources and produce graduates who have mastered skills such as "Chemical Literature Skills" (American Chemical Society Committee on Professional Training (ACS-CPT), 2008). These skills include effectively evaluating the peer reviewed literature and using specialized database-searching tools, and correspond directly to the ALA/ACRL/STS Task Force on Information Literacy for Science and Technology [STS-TFILST]

(2006) Information Literacy Standards for Science and Engineering/Technology (hereafter, Standards) (Figure 4.1). The ACS-CPT guidelines state that "Approved programs must provide instruction on the effective retrieval and use of the chemical literature" (ACS-CPT, 2008, p. 14). These guidelines can serve as an entry point for conversations with faculty, which may lead to opportunities to incorporate information literacy into the curriculum.

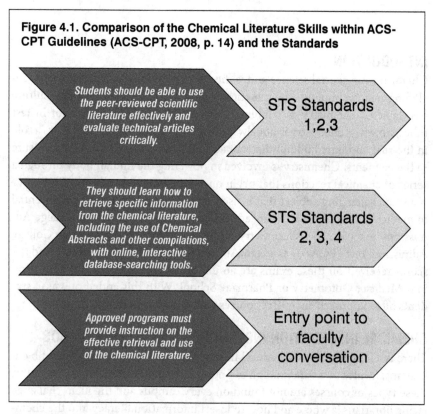

Figure 4.1. Comparison of the Chemical Literature Skills within ACS-CPT Guidelines (ACS-CPT, 2008, p. 14) and the Standards

Students should be able to use the peer-reviewed scientific literature effectively and evaluate technical articles critically.

STS Standards 1,2,3

They should learn how to retrieve specific information from the chemical literature, including the use of Chemical Abstracts and other compilations, with online, interactive database-searching tools.

STS Standards 2, 3, 4

Approved programs must provide instruction on the effective retrieval and use of the chemical literature.

Entry point to faculty conversation

The ACS Division of Chemical Information (CINF) developed *Chemical Information Retrieval*, a companion document which outlines specific topics and skills in which ACS-certified degree holders should be proficient (American Chemical Society Division of Chemical Information [ACS-CINF], 2008). The document states that ACS-certified graduates should have a demonstrable understanding of specific resources such as *Chemical Abstracts* and the *Handbook of Chemistry and Physics*, and graduates should be able to perform skills such as locating chemical and physical properties of compounds. *Chemical Information Retrieval* provides concrete chemistry performance indicators that correspond to the content within the Standards (Figure 4.2).

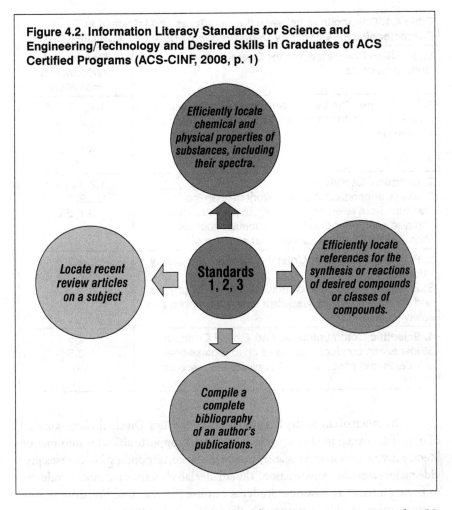

Figure 4.2. Information Literacy Standards for Science and Engineering/Technology and Desired Skills in Graduates of ACS Certified Programs (ACS-CINF, 2008, p. 1)

The Special Libraries Association (SLA), Chemistry Division and ACS CINF Education committee (2011) have outlined skills and undergraduate proficiencies from the Standards and the ACS CPT Guidelines in *Information Competencies for Chemistry Undergraduates: The Elements of Information Literacy* (hereafter, Information Competencies). This is an essential resource for librarians who work with chemistry undergraduates. The main objective of the document is to help facilitate the integration of information literacy into the chemistry curriculum and highlight recommended resources that were selected by librarians from the SLA Chemistry division and the ACS CINF Education Committee. A mapping of the Standards to the Information Competencies is shown in Table 4.1.

Table 4.1. Connections between the Standards and Information Competencies for Chemistry Undergraduates (SLA and ACS CINF, 2011)	
Information Competencies for Chemistry Undergraduates	**Standards & Performance Indicators**
1. Big Picture: The Library and Scientific Literature Students understand the organization of the library and scientific literature.	1.2, 1.3, 1.4 2.2, 2.3 3.1, 3.2 4.1, 4.2 5
2. Chemical Literature Students understand the framework of chemical literature, from reference materials to patents. Students can also search for chemical information such as chemical structures, reactions, and syntheses.	1.2, 1.3, 1.4 2.1, 2.2, 2.3, 2.4 3.1, 3.2
3. Properties, Spectra, Crystallographic, and Safety Information Students can search the chemical literature for physical and chemical properties, spectra, crystallographic, and safety information.	1.2, 1.3, 1.4 2.1, 2.2, 2.3, 2.4 3.1, 3.2
4. Scientific Communication and Ethical Conduct Students can conduct their research to professional standards and effectively communicate their research to others.	2.5 3.5 4 5

Each branch of chemistry has unique characteristics. Due to the large amount of material covered in chemistry lecture courses, opportunities for information literacy instruction may be scarce; however, the corresponding lab courses provide major gateways to instruction. The smaller lab class sizes are more conducive to having interactive sessions. Students can use the chemical literature to support their work and investigations in the lab. For example, students would need to know the chemical properties of several chemical compounds to make the correct determination of their unknown compound. This chapter will provide examples of information literacy activities for specific chemistry courses.

NAVIGATING THE CHEMICAL LITERATURE

The language of chemistry can be complex, and there are a number of things to emphasize to students as they begin their research. Chemical compounds can have multiple names: a common name (e.g., ibuprofen) that is often known to the layperson, a trade name (e.g., Advil) selected by its manufacturer, and a name determined by International Union of Pure and Applied Chemistry (IUPAC) (e.g., 2-[4-(2-methylpropyl)phenyl]propanoic acid). Chemical Abstracts Service (CAS) assigns Registry Numbers (CAS RN) to compounds for identification as well. Although compounds can share molecular formulas, the elements of each compound can be arranged differently resulting in different chemical structures. These structures are depicted by two-dimensional chemical drawings or three-dimensional models. They can also be determined by the IUPAC International Chemical Identifier (InCHi) or simplified molecular-input line-entry system (SMILES). Physical properties of compounds can be difficult to find. Depending on the property, the information could be found easily in a table within reference materials or buried in the journal literature. Free web based tools such as *NIST Chemistry WebBook, ChemSpider* and *ChemID Plus* can help students get started with finding the chemical information they need.

To search the literature effectively, students must understand the chemistry literature publication cycle, from the lab notebook to reference materials and review articles. In Section 2 "Chemical Literature" of the Information Competencies the following key recommended resources are outlined:

- Background Information (e.g., *Hawley's Condensed Chemical Dictionary* and *Kirk-Othmer Encyclopedia of Chemical Technology*);
- Articles and Other Chemical Literature (e.g., *SciFinder* and *Web of Science*);
- Patents (e.g., *Reaxys, SciFinder,* and United State Patent and Trademark Office–USPTO,);
- Chemical Substances, Reactions, and Synthesis (e.g., *ChemSpider, Combined Chemical Dictionary, Merck Index, Reaxys,* and *SciFinder*).

For lower division students, contrasting multidisciplinary databases such as *Academic Search Premier* and *Web of Science* with specialized science resources such as *BIOSIS Previews* and *Knovel* will show students the variety of information resources and differences in organization. For Chemistry majors and upper division students, chemistry-centric databases such as *SciFinder* and *Reaxys* will illustrate the importance of specialized tools for research, especially when searching for information on synthesis and physical properties. In these specialized tools, students can search the database for articles, synthesis and property information as well as draw or import their structures.

Examples of key activities to use as information literacy building blocks at appropriate levels within the chemistry curriculum are presented in Figure 4.3. Building relationships with faculty is a major component to successfully incorporating the information literacy activities into the curriculum. Many faculty members choose to introduce and demonstrate the scientific databases themselves; however, some faculty work with their librarians to incorporate information literacy in their courses. At some institutions, general and organic chemistry have too many sections for sustainable library instruction. In these situations, the upper division courses can have more meaningful assignments and would be better sections to target. The remainder of this chapter highlights the characteristics of courses within the chemistry sequence and provides examples of information literacy activities to incorporate along the way.

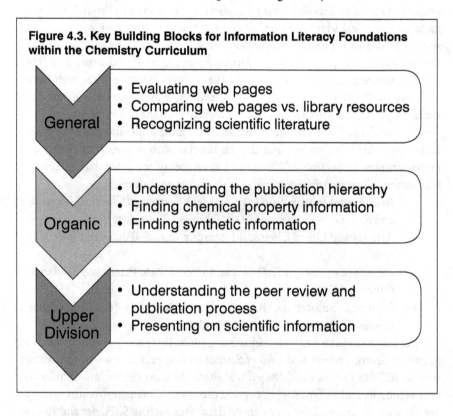

Figure 4.3. Key Building Blocks for Information Literacy Foundations within the Chemistry Curriculum

General
- Evaluating web pages
- Comparing web pages vs. library resources
- Recognizing scientific literature

Organic
- Understanding the publication hierarchy
- Finding chemical property information
- Finding synthetic information

Upper Division
- Understanding the peer review and publication process
- Presenting on scientific information

GENERAL CHEMISTRY

First year chemistry courses are survey courses, which provide science and engineering majors with an overview of chemistry beginning with the principles of measurement to rates of chemical reactions. General chemistry students will learn about the periodic table, nomenclature and chemical reactions. In non-major chemistry courses (typically a course on environmental chemistry or a "chemistry and society" course), students are taught an overview of chemistry with a lens focused on the environment and the impact chemistry has had on society and the environment. These courses provide the foundation for scientific concepts rooted in accessible examples seen in everyday life. They often fulfill the general education science requirement and may be writing-intensive courses as well. The corresponding general chemistry and non-major chemistry lab courses provide the foundation for lab techniques. Although the majority of the students enrolled in these courses are not science majors, the courses provide opportunities to build working relationships with chemistry faculty and instructors. Use the following examples to create opportunities for incorporating information literacy into general chemistry and non-major chemistry courses.

Safety in the Lab

General chemistry lab students are expected to learn and practice the safe handling of chemicals. An important skill is locating reliable Material Safety Data Sheet (MSDS) information on the chemicals to be used in the lab. Librarians can provide assistance to instructors and faculty by creating instruction sessions for finding MSDS information and finding basic chemical information within the library and online. The Information Competencies Section 3.4, "Safety Information" recommends resources such as *CRC Handbook of Laboratory Safety*, *TOXNET* and links to several MSDS collections.

Knowing the physical properties of the chemicals used in the lab will help students know how to handle the chemicals in a safe manner. Resources such as the *Merck Index* and the *CRC Handbook of Chemistry and Physics* also provide important physical property information, including physical descriptions, boiling points, melting points, and solubility information. This information will be used to identify the correct compounds, handle the chemicals in a safe manner and complete the general chemistry lab activities.

Science on the Web and Current Topics in Science/Chemistry

Consumers are bombarded by websites purporting the latest scientific claim. Information evaluation is the crux of information literacy. Librarians can approach general chemistry and non-major chemistry instructors with activities that evaluate science found on websites and discussed in the latest news reports. Website evaluations can be used as a springboard for class discussion. Using satirical websites such as *Dihydrogen Monoxide Research Division* (http://dhmo.org) and *Havidol* (http://havidol.com/), librarians can point out the importance of evaluating information found on the web. These examples can be used in comparison with websites produced by the U.S. Environmental Protection Agency (http://www.epa.gov/climatechange/) and websites for pharmaceutical companies. During the course of the library session, students will determine the validity of the websites and develop the skills for evaluating resources. These skills can also be translated into evaluating the literature on the latest science news headlines. Students can explore science topics on the web and compare the results found in scientific databases such as *Web of Science* and *BIOSIS Previews*.

Gawalt and Adams (2011) have incorporated an inquiry based approach to investigating the chemical literature into an invitation-only honors program for first semester general chemistry students to achieve objectives listed in the Standards, specifically Standard 2. In this program, the librarians meet with the students a total of seven times: three library meetings (introduction to library navigation and specialized library resources) and four classroom meetings (introduction to primary literature and critical thinking). The librarians give short presentations highlighting databases and aspects of chemistry literature. The students investigate the databases while completing structured assignments and engaging in discussion focused on reading selected journal articles. Pieces of this approach can be used to kick start an information literacy program in General Chemistry.

ORGANIC CHEMISTRY

Organic chemistry is usually the second full year course in the undergraduate chemistry sequence. The student's performance in this subject typically determines if s/he will continue along the medical/pre-professional track, as it is required for admission into Medical, Pharmacy, Dental and Veterinary programs (https://www.aamc.org/students/applying/requirements/).

Organic chemistry is focused on carbon based compounds and their structure and reactivity, and students quickly become immersed in nomenclature and reactions. In this course, students become even more focused on synthesis of compounds, and they develop an understanding of how compounds are made and transformed through their chemical bonds. Students study reaction mechanisms, the step by step process in which chemical bonds are broken and created. This course provides an entry point for several information literacy opportunities from finding physical properties to the history behind named reactions to the synthesis of specific compounds.

Physical Properties

Finding physical properties are important pre-lab activities for organic chemistry. Students will need the physical description, molecular weight, melting points, and boiling points of the compounds to be used in the lab. Molecular weights are used to calculate the amount of materials needed prior to the beginning of lab and to determine the yield. Melting points are used to determine the purity of the synthesized compounds once the exercise is completed. This information can be found in the *CRC Handbook of Chemistry and Physics, Knovel Critical Tables, Merck Index, NIST Chemistry WebBook* and other resources listed in Section 3.1 "Properties" in the Information Competencies.

Named Reactions and Research Compound Projects

Students can gain experience navigating the chemical literature while investigating a specific compound or while researching the history of a "Named Reaction." In the "Named Reactions" paper (a typical assignment) students will research the origins of the named reaction by finding the original literature citations. As students learn how to search within the chemical literature for this project, they will achieve many of the skills highlighted within the ACS "Chemical Information Retrieval" document (ACS CINF, 2008).

Organic chemistry instructors in the department may already be assigning a project in which their students research a specific compound. While researching this compound, students will need to incorporate nomenclature (including IUPAC, common and trade names), molecular formula, molecular structure and the compound's Chemical Abstract Registry Number (CAS RN) into their search strategies. Instructors may also require students to include the historical background and significance of the compound in medicine or industry as well as the synthesis of the compound from reasonable starting materials. Students will use a wide variety of resources including *Knovel, Merck Index, Kirk Othmer Encyclopedia of Chemical Technology, PubMed, Web of Science,* and *SciFinder* to find information on their compounds. In addition, they will often encounter the patent literature for the first time during this project.

Patents provide the bridge between synthetic reactions learned in the classroom and synthetic processes utilized in industry. MacMillan and Shaw (2008) outline methods to introduce patents to second year students enrolled in an Industrial Organic Chemistry Course. Students use the USTPO search interface to locate several patents for their assigned industrial chemicals. By evaluating the patents for their compounds, students are able to determine established and mature chemical technology versus currently competitive processes. Chapter 6 of this book provides additional information about patents and how to integrate them into the curriculum.

When researching a specific compound, students may also research the mechanism of the reaction and determine if their selected compound is involved in the synthesis of other important molecules. Applicable resources for Named Reaction and Research Compound activities are outlined in Section 2 "Chemical Literature" in Information Competencies. Several librarians have created library and information literacy activities that take organic chemistry students "beyond Google" (Peters, 2011) and situate the "Librarian in the Laboratory" (Tomaszewski, 2011). These activities and more are highlighted in Table 4.2 at the end of this chapter.

UPPER DIVISION COURSES

General chemistry and organic chemistry are the foundations for the upper division courses. In upper division chemistry courses, the students are focused on specific experiments, and they may be starting their own individual research projects. Instructor engagement is a crucial element for successfully incorporating information literacy into the curriculum, particularly at this level. Librarians should target faculty who champion information literacy and tailor the library sessions to the type of information that lab activity is requiring.

Analytical Chemistry and Instrumental Analysis

Analytical chemistry and Instrumental Analysis is focused on solving the mysteries, "What is this made of?" and "How much is in there?" Students will be making determinations about chemical compositions and structures through a variety of methods including infrared (IR) and ultraviolet (UV) spectroscopic analysis, Nuclear Magnetic Resonance (NMR), and Mass Spectrometry (MS). Advanced classes in Instrumental Analysis take an in-depth look at the instrumentation and calculations used to make these determinations (Lewis, 2007).

Biochemistry

This is the study of all aspects of chemistry which pertain to living organisms was once considered merely a subset of Chemistry and now an independent area of science (Lewis, 2007). Analytical, physical and inorganic chemistry each play a big role in characterization and reactions within biochemistry. Many pre-professional students take this course as an elective to build a foundation for future courses in nutrition, pharmacology, medicinal chemistry and toxicology.

Geochemistry

Geochemistry is the study of the chemical composition, physiochemical and geological processes that produce and modify minerals and rocks (Lewis, 2007). Environmental topics such as biosequestration, radioactive waste disposal, and analysis of groundwater are covered in this course.

Inorganic Chemistry

This is the branch of chemistry which does not deal with hydrocarbons and their derivatives. Topics include atomic structure, crystallography, chemical bonding, catalysis, coordination compounds, ceramics and electrochemistry (Lewis, 2007).

Physical Chemistry

Physical chemistry is focused on "How things work?" on a molecular level. Using examples from other chemical areas, physical chemistry applies the laws of physics to the chemical phenomena (Lewis, 2007). In upper division courses, students will be expected to calculate constants based on their lab experiences and verify these calculations within the supporting literature.

Science and Chemistry specific databases such as *SciFinder and Web of Science* can be used as springboards to encourage lifelong learning; use these databases to search and "shop" for potential research advisors and graduate programs during an instruction session and make lasting impressions on forward thinking students. *Knovel* contains reference materials that can bridge theory and applications of Analytical and Instrumental Methods in Medicine and Industry. For Geochemistry courses, focus on databases that allow searching by geographic location, such as *GeoRef*. For Inorganic courses, teach students to find crystallographic data in resources such as the *International Union of Crystallography*. For Biochemistry and Physical chemistry courses, students may need to use a wide variety of resources, including *Knovel, PubMed, SciFinder, and Web of Science* to find the physical constants they are seeking. Additional resources for these upper division courses are listed in Section 3 "Properties, Spectra, Crystallographic and Safety Information" in the Information Competencies (SLA and ACS CINF, 2011).

DIRECTED STUDY, UNDERGRADUATE RESEARCH OR INTERNSHIPS

Once the foundational chemistry courses have been taken, many students begin their undergraduate research experiences. Some highly motivated students begin their research careers in their first or second year of study. Target "Directed Study" or "Internship" courses, and create workshops geared toward students presenting their research during on-campus events, regional and national meetings. Topics to consider are poster creation and presentation skills. Section 4 "Scientific Communication and Ethical Conduct" in the Information Competencies covers writing styles, citation and reference manager software and poster presentation resources that the librarian can use in classes on these topics (SLA and ACS CINF, 2011).

RESOURCES FOR LIBRARIANS

The Chemistry Librarian community is a vibrant and engaging community which has developed many resources to help colleagues without formal education or experience in chemistry. Some recommended resources important to chemistry are listed in Appendix 4.1 with a few highlighted below. *Chemical Information Sources* is a comprehensive wiki, which thoroughly explains aspects of chemical information resources from "How and Where to Start" to "How and Where to Search." The mechanics and details of "Analytical Chemistry Searches," "Physical Property Searches" and more are explained within this resource. *Chemical Information Sources* provides the background and foundation for the associated *Chemical Information Instruction Materials (CIIM)* and *Selected Internet Resources for Chemistry (SIRCh)*. CIIM and SIRCh highlight and link to important print and web resources in Chemistry. Resources listed include *CAS Source Index (CASSI)* for deciphering and confirming abbreviated publication titles and ISSN codes and the *Spectral Database for Organic Compounds (SDBS)*. These wikis were originally lecture notes and texts created by Garry Wiggins, Librarian Emeritus at Indiana University and are now available for the chemistry librarian community to contribute to and update as needed.

XCITR (eXplore Chemical Information Teaching Resources) is a repository of instruction material focused on chemical information and is a collaborative project between the Computer-Information-Chemistry (CIC) division of the German Chemical Society and ACS—CINF and is hosted by FIZ Chemie Berlin. Materials are searchable and are also organized by "subject", "subject category" and "resources covered."

Professional library organizations, including the SLA Chemistry Division, offer "Chemistry for the non-chemist" pre-conference workshops during annual meetings. These workshops provide a hands-on introduction to the basic principles of chemistry and approaches to addressing the research needs of chemists. Occasionally the workshops have also been offered as webinars (see Appendix 4.1).

Table 4.2. Chemical Information Literacy Activities Cited in the Literature			
In this activity students will...	Standard and Performance Indicator*	Area of Chemistry	Reference
Use an inquiry based approach to find chemical terms and substance properties from chemically related primary sources and library information.	1, 2	General	Gawalt & Adams, 2011
Navigate the library catalog for books, articles, and conference proceedings. Using subscription-based resources (The Merck Index and CHEMnetBASE), students find basic chemical information.	1.2, 1.3, 1.4, 2.1, 2.2, 2.3, 2.4	Organic	Peters, 2011

Table 4.2. Chemical Information Literacy Activities Cited in the Literature

In this activity students will...	Standard and Performance Indicator*	Area of Chemistry	Reference
Use CHEMnetBASE and Organic Syntheses to find chemical and synthetic information and original literature on an organic compound.	1.2, 1.3, 1.4, 2.1, 2.2, 2.3, 2.4, 2.5, 3.1	Organic	Peters, 2011
Use Reaxys, Science of Synthesis and SciFinder® Web to find chemical and synthetic information on their laboratory compounds.	1.2, 1.3, 1.4, 2.1, 2.2, 2.3, 2.4, 2.5, 3.1, 3.3, 3.4, 4.1, 4.2	Organic	Peters, 2011
Prepare derivatives, characterize their products using melting points, IR, 1H NMR and 13C NMR; and calculate their percent yield. The characterizing information and yield is compared to literature values obtained in the MDL CrossFire (Beilstein) database. Students are also expected to find the original literature.	1.2, 1.3, 1.4, 2.1, 2.2, 2.3, 2.4, 3.1	Organic	Tomaszewski, 2011
Prepare an annotated bibliography of a pharmaceutical or well-known chemical.	1.2, 1.3, 1.4, 2.1, 2.2, 2.3, 2.4, 2.5	Organic	Jensen, Narske & Ghinazzi, 2010
Analyze the main contents of a Journal of Organic Chemistry article	1.2, 1.3, 1.4, 3.1, 3.3	Organic	Jensen, Narske & Ghinazzi, 2010
Conduct a literature search of a named reaction and prepare a bibliography.	1.2, 1.3, 1.4, 2.1, 2.2, 2.3, 2.4, 2.5, 3.1, 3.2, 3.3	Organic	Jensen, Narske & Ghinazzi, 2010
Critically evaluate biochemistry information sources and use the library's resources to find articles on biochemical topics.	1 & 2	Biochemistry	Zhang, 2007

Table 4.2. Chemical Information Literacy Activities Cited in the Literature

In this activity students will...	Standard and Performance Indicator*	Area of Chemistry	Reference
Use the chemical literature to research and "Design your own disease."	1, 2, 3, 4, 5	Biochemistry	Flynn, 2010
Research Material Safety Data Sheet information.	1, 2, 3	Analytical	Walczak & Jackson, 2007
Plan attendance at an upcoming professional meeting.	1, 2, 4, 5	Analytical	Walczak & Jackson, 2007
Prepare and invite a seminar speaker.	3, 4, 5	Analytical	Walczak & Jackson, 2007
Select an analytical chemistry topic and perform a literature review.	2, 3, 4	Analytical	Walczak & Jackson, 2007
Participate in the peer-review process by creating and reviewing a bibliography and a manuscript.	4.3, 4.4, 4.5, 4.6	Analytical	Walczak & Jackson, 2007
Conduct a literature review and annotated bibliography on a new instrument concept or product in a specific area.	1, 2, 3, 5	Analytical	Henderson, 2010
Present a proposal on their new instrument concept to their peers for evaluation.	1, 2, 3	Analytical	Henderson, 2010

* (STS-TFILST, 2006)

REFERENCES

ALA/ACRL/STS Task Force on Information Literacy for Science and Technology [STS-TFILST]. (2006). Information literacy standards for science and engineering/technology. Retrieved from http://www.ala.org/acrl/standards/infolitscitech

American Chemical Society Committee on Professional Training (ACS-CPT). (2008). Undergraduate professional education in chemistry. Retrieved from http://portal.acs.org/portal/PublicWebSite/about/governance/committees/training/acsapproved/degreeprogram/WPCP_008491

American Chemical Society Division of Chemical Information (ACS-CINF). (2008). Chemical information retrieval. Retrieved from http://portal.acs.org/portal/PublicWebSite/about/governance/committees/training/acsapproved/degreeprogram/CTP_005584

Flynn, N. (2010). "Design your own disease" assignment: Teaching students to apply metabolic pathways. Journal of Chemical Education, 87, 799–802. doi:10.1021/ed100279z

Garritano, J. R., & Culp, F. B. (2010). Chemical information instruction in academe: Who is leading the charge? Journal of Chemical Education, 87, 340–344. doi:10.1021/ed800085h

Gawalt, E. S., & Adams, B. (2011). A chemical information literacy program for first-year Students. Journal of Chemical Education. 88, 402–407. doi: 10.1021/ed100625n

Henderson, D. E. (2010). A chemical instrumentation game for teaching critical thinking and information literacy in instrumental analysis courses. Journal of Chemical Education, 87, 412–415. doi:10.1021/ed800110f

Jensen, D., Narske, R., & Ghinazzi, C. (2010). Beyond chemical literature: Developing skills for chemical research literacy. Journal of Chemical Education, 87, 700–702. doi:10.1021/ed1002674

Lewis Sr., R. J. (2007). Hawley's condensed chemical dictionary (15th ed.). Hoboken, NJ: John Wiley & Sons. Retrieved from http://www.knovel.com

MacMillan, M., & Shaw, L. (2008). Teaching chemistry students how to use patent databases and glean patent information. Journal of Chemical Education, 85, 997. doi:10.1021/ed085p997

Peters, M. (2011). Beyond Google: Integrating chemical information into the undergraduate chemistry and biochemistry curriculum. Science & Technology Libraries, 30, 80–88. doi:10.1080/0194262X.2011.545671

Special Libraries Association Chemistry Division and American Chemical Society Division of Chemical Information. (2011). Information competencies for chemistry undergraduates: The elements of information literacy (2nd ed.). Retrieved from http://units.sla.org/division/dche/il/cheminfolit.pdf

Tomaszewski, R. (2011). A science librarian in the laboratory: A case study. Journal of Chemical Education. 88, 755. doi:10.1021/ed1000735

Walczak, M. M., & Jackson, P. T. (2007). Incorporating information literacy skills into analytical chemistry: An evolutionary step. Journal of Chemical Education, 84, 1385–1390.

Zhang, L. (2007). Promoting critical thinking, and information instruction in a biochemistry course. Issues in Science & Technology Librarianship, 51. Retrieved from http://www.istl.org/07-summer/refereed.html

APPENDIX 4.1. RECOMMENDED RESOURCES

Database	Access	URL
Chemical Abstracts Services Source Index (CASSI)	Open access	http://cassi.cas.org/search.jsp
Chemical Information Sources	Open access	http://en.wikibooks.org/wiki/ Chemical_Information_Sources
Chemical Information Instruction Materials	Open access	http://en.wikibooks.org/wiki/ Chemical_Information_Sources/ CIIM
ChemID Plus	Open access	http://chem.sis.nlm.nih.gov/ chemidplus/
ChemSpider	Open access	http://www.chemspider.com/
Combined Chemical Dictionary	Subscription-based	http://www.chemnetbase.com/
CRC Handbook of Chemistry and Physics	Subscription-based	http://hbcpnetbase.com/
CRC Handbook of Laboratory Safety	Subscription-based	http://www.crcnetbase.com/ isbn/978-0-8493-2523-6
Hawley's Condensed Chemical Dictionary	Subscription-based	http://onlinelibrary.wiley.com/ book/10.1002/9780470114735
International Union of Crystallography	Subscription-based	http://it.iucr.org/
Kirk-Othmer Encyclopedia of Chemical Technology	Subscription-based	http://onlinelibrary.wiley.com/ book/10.1002/0471238961
Knovel	Subscription-based	http://www.knovel.com
The Merck Index	Subscription-based	http://www.merckbooks.com/ mindex/
MEDLINE/PubMed	Open access (not full text)	http://www.ncbi.nlm.nih.gov/ pubmed
NIST Chemistry WebBook	Open access	http://webbook.nist.gov/ chemistry/
Reaxys	Subscription-based	https://www.reaxys.com/info/
Selected Internet Resources for Chemistry (SIRCh)	Open access	http://en.wikibooks.org/wiki/ Chemical_Information_Sources/ SIRCh
SciFinder	Subscription-based	http://www.cas.org/products/ sfacad/index.html

APPENDIX 4.1. RECOMMENDED RESOURCES

Database	Access	URL
SLA Division of Chemistry "Chem Info for the Non-Practitioner" webcast recording and slides	Open access	http://chemistry.sla.org/2010/webcast-recording-slides-chem-info-for-the-non-practitioner/
Spectral Databases for Organic Compounds (SDBS)	Open access	http://riodb01.ibase.aist.go.jp/sdbs/cgi-bin/cre_index.cgi?lang=eng
TOXNET	Open access	http://toxnet.nlm.nih.gov/
Web of Science	Subscription-based	http://thomsonreuters.com/products_services/science/science_products/a-z/web_of_science/
xCITR eXplore Chemical Information Teaching Resources	Open access	http://www.xcitr.org/

IMPORTANT INFORMATION LITERACY STANDARDS FOR HUMAN NUTRITION

2. The information literate student acquires needed information effectively and efficiently.
   1. Selects the most appropriate investigative methods or information retrieval systems for accessing the needed information.
   2. Constructs and implements effectively designed search strategies.
   3. Retrieves information using a variety of methods.
   4. Refines the search strategy if necessary.
   5. Extracts, records, transfers, and manages the information and its sources.

3. The information literate student critically evaluates the procured information and its sources, and as a result, decides whether or not to modify the initial query and/or seek additional sources and whether to develop a new research process.
   2. Selects information by articulating and applying criteria for evaluating both the information and its sources.
   3. Synthesizes main ideas to construct new concepts.
   4. Compares new knowledge with prior knowledge to determine the value added, contradictions, or other unique characteristics of the information.
   5. Validates understanding and interpretation of the information through discourse with other individuals, small groups or teams, subject-area experts, and/or practitioners.
   7. Evaluates the procured information and the entire process.

4. The information literate student understands the economic, ethical, legal, and social issues surrounding the use of information and its technologies and either as an individual or as a member of a group, uses information effectively, ethically, and legally to accomplish a specific purpose.
   1. Understands many of the ethical, legal and socio-economic issues surrounding information and information technology.
   3. Acknowledges the use of information sources in communicating the product or performance.
   6. Communicates the product or performance effectively to others.

HUMAN NUTRITION

CHAPTER 5

Rebecca K. Miller
College Librarian for Science, Life Sciences, and Engineering
Virginia Tech

FROM DOMESTIC SCIENCE TO SCIENTIFIC PROGRESS: HUMAN NUTRITION IN CONTEXT

The academic and professional study of human nutrition and related topics emerged from programs in home economics that grew in popularity in the United States from the 1870's onward and that began focusing on the scientific study of nutrition (Stage, 1997). By the early twentieth century, the "domestic scientists" graduating from these home economics programs started developing new professional roles for themselves in food management, as hospital dietitians, and as agents in the Cooperative Extension System (Nyhart, 1997; Babbitt, 1997). Today, most major research universities house nutrition-related departments and programs that support ambitious research agendas embracing a wide range of topics. Various aspects of health, obesity, genetics, chronic disease, behavioral and community interventions, physical activity and exercise, food studies, and effective health communication all represent research areas that may be located in nutrition programs across the nation.

In addition to participating in collaborative interactions to advance research, nutrition-related programs in U.S. universities endeavor to prepare their undergraduate and graduate students for an assortment of professional roles and occupations. Students choosing to study nutrition, food, and exercise often aspire to careers in dietetics, hospitality and food industries, fitness promotion and sports medicine, and various healthcare fields such as public health, physical therapy, and veterinary care. Although students of human nutrition and related disciplines have many options available to them after finishing their programs of study, most nutrition departments or programs will emphasize one or two curricula that lead directly to professional preparation and certification. Departmental emphasis will vary among institutions; however, most nutrition departments will include dietetic programs accredited by the Academy of Nutrition and Dietetics (formerly the American Dietetic Association; hereafter, Academy). Other professional certification programs or curricula that nutrition departments may work with or support include the American College of Sports Medicine's (ACSM) Certified Health Fitness Specialist program, the

National Athletic Trainers' Association (NATA) Certified Athletic Trainer program, various types of health or nutrition "coach" certification programs, and even educational licensing programs. Preparation for the specialized area of dietetics, though, remains the most pervasive form of professional training within nutrition programs at universities across the United States. This chapter will align program accreditation and curriculum standards developed by the Academy with Information Literacy Standards for Science and Engineering/ Technology (hereafter, Standards, (ALA/ACRL/STS Task Force on Information Literacy for Science and Technology [STS-TFILST], 2006) in order to highlight particular areas where librarians can use information literacy principles to advance the disciplinary goals of nutrition programs in American colleges and universities.

The Academy regulates and provides standardization for food and nutrition professionals in the United States. The professional roles of Registered Dietitian (RD) and Dietetic Technician (DT) require at least the successful completion of an Academy-accredited program of study; RD preparation also includes successfully completing an Academy-accredited Dietetic Internship program and passing an exam. The term *nutritionist* is not regulated by the Academy, and does not specify the type of training or coursework completed by a person using that label, although some states have licensure laws that define the scope of practice for a person using the designation *nutritionist*. Even so, many university departments or programs relating to nutrition do maintain the term *nutrition* in their names. This explanation of professional standardization serves to illustrate the rigorous guidelines and regulations that shape the core curricula of dietetics in many nutrition-oriented programs at colleges and universities in the United States. *Dietetics* refers to the area of knowledge surrounding scientific study of nutrition and its practical applications, and is specifically concerned with diet and its effects on health. The Academy's Accreditation Council for Education in Nutrition and Dietetics (ACEND, formerly the Commission on Accreditation for Dietetics Education) maintains high standards for students graduating from Academy-accredited programs, and identifies specific competencies that accredited programs must embrace. ACEND produces guidelines detailing the foundational knowledge and competencies for dietitian education; as of the writing of this chapter, these guidelines were most recently updated in 2008 and emphasize the importance of understanding the nature of scientific information and research in the electronic environment (Accreditation Council for Education in Nutrition and Dietetics [ACEND], 2008). Not surprisingly, several of the outcomes described in the *ACEND 2008 Foundation Knowledge and Competencies for Dietitian Education* (hereafter *ACEND Guidelines*; see Appendix 5.1) match those from the Standards.

ACEND MEETS ACRL: EFFECTIVELY TRANSLATING INFORMATION LITERACY CONCEPTS

Although information literacy may be "common to all disciplines," the language used to describe it is not (Association of College & Research Libraries, 2000). Effectively communicating the importance of incorporating information literacy standards into various curricula may be difficult, but not when some of the information literacy standards and performance indicators can be translated into terminology already used by a particular discipline. The ACEND Guidelines provide this disciplinary vocabulary that academic librarians can use to stress the importance of information literacy, demonstrate an understanding of the field of dietetics and nutrition, and provide genuine value to nutrition departments by helping them meet ACEND accreditation standards.

The ACEND Guidelines include four specific Knowledge Requirements (KR) that are broken down into detailed learning outcomes, similar to the Standards. According to these Knowledge Requirements, Standards 2, 3, and 4 represent the most imperative skill sets to address in instruction sessions with nutrition students, given the curricular guidelines and standards developed by ACEND.

Table 5.1. Comparison of Relevant Sci/Tech Info Lit (STS-TFILST, 2006) & ACEND Knowledge Requirements (ACEND, 2008)		
ACRL Sci/Tech Information Literacy Standards	**ACEND Knowledge Requirements (KR)**	**Shared Performance Indicators/Learning Outcomes**
Standard 2: The information literate student acquires needed information effectively and efficiently.	**KR 1.1.a:** Students are able to demonstrate how to locate, interpret, evaluate, and use professional literature to make ethical evidence-based practice decisions. **KR 1.1.b:** Students are able to use current information technologies to locate and apply evidence-based guidelines and protocols.	Locate appropriate resources for searching for information. Develop appropriate search strategies and reevaluate those strategies when necessary. Gain confidence in using new technologies for discovering, and contributing to, new research.

Table 5.1. Comparison of Relevant Sci/Tech Info Lit (STS-TFILST, 2006) & ACEND Knowledge Requirements (ACEND, 2008)		
ACRL Sci/Tech Information Literacy Standards	**ACEND Knowledge Requirements (KR)**	**Shared Performance Indicators/Learning Outcomes**
Standard 3: The information literate student critically evaluates the procured information and its sources...	**KR 1.1a** **KR 1.1b** (See above)	Be able to critically evaluate the research and information found. Be able to interpret and apply the information found through research. Understand appropriate research practices and methodologies; be able to recognize and use these appropriate strategies when reading and contributing to research.
Standard 4: The information literate student understands the economic, ethical, legal, and social issues surrounding the use of information and its technologies and either as an individual or as a member of a group, uses information effectively, ethically, and legally to accomplish a specific purpose.	KR 2.1.a: Students are able to demonstrate effective and professional oral and written communication and documentation and use of current information technologies when communicating with individuals, groups, and the public.	Use effective communication methods to communicate with target groups or individuals. Document and acknowledge the use of information sources in written, and other, communications. Critically evaluate all communications developed in order to ensure effectiveness of the presentation of information.

ACEND defines KR1, Scientific and Evidence Base of Practice, as the "integration of scientific information and research into practice" (ACEND, 2008, p. 1). With two specific learning outcomes, this knowledge requirement underscores students' ability to "locate, interpret, evaluate and use professional literature" and to "use current information technologies to locate and apply evidence-based guidelines and protocols" (ACEND, 2008, p. 1). The spirit of these learning outcomes remains conceptually very close to the performance indicators described in Standards 2 and 3. Acknowledging, respectively, that the information literate student "acquires needed information effectively and efficiently" and "critically evaluates the procured information and its sources," Standards 2 and 3 can be applied at a variety of levels and depths in the classroom.

Next, the ACEND KR2, Professional Practice Expectations, stipulates that students will "develop a variety of communication skills" (ACEND, 2008, p. 2). One of the two specific learning outcomes associated with KR 2 asserts that students graduating from Academy-accredited programs should be able to "demonstrate effective and professional oral and written communication and documentation and use...current information technologies when communicating" (ACEND, 2008, p. 2). Fundamentally, these learning outcomes mirror Standard 4 and its performance indicators, addressing a student's ability to "acknowledge the use of information sources in communicating," apply information creatively, and communicate effectively with others (STS-TFILST, 2006). Librarians have a significant role to play in assisting students in nutrition programs gain the skills to communicate effectively, ethically, and legally, and can use a wide variety of instructional strategies to help students continue to build upon these skills at all levels.

Librarians supporting nutrition programs can add the most value to the curricula through instruction related to Standards 2, 3, and 4, which, as noted above, correlate with KR1 and KR2 of the *ACEND Guidelines*. Librarians should leverage the language used in these Knowledge Requirements in discussions and collaborations with faculty and students. In doing so, the librarian situates himself or herself directly within the discipline, creating a connection so that information literacy instruction sessions can be perceived as directly applicable to the objectives of specific courses as well as to the professional goals of the overall curriculum.

LIBRARIAN MEETS STUDENTS: TRANSLATING STANDARDS INTO INSTRUCTION

While the *ACEND Guidelines* assist librarians with translating information literacy standards into a language intelligible to faculty and students in nutrition programs, effectively translating information literacy concepts and standards into instruction sessions, lesson plans, and learning activities can vary greatly depending on types and levels of students with which librarians may be working. As always, requesting to view a syllabus and description of course assignments before planning instructional session(s) for a particular class can be helpful in determining how a particular instructor plans to address the overall objectives described in the *ACEND Guidelines*. By taking into consideration the larger objectives of the course and the curriculum, library instruction sessions can maximize the impact and value of information literacy concepts to the nutrition curriculum. This section of the chapter explores strategies for applying Standards 2, 3, and 4 in different types and levels of nutrition courses. Keep in mind that, although this chapter organizes instructional strategies by the information literacy standard that they relate to, a good information literacy session may include focusing on more than one standard within a single session. For this reason, consider the ideas discussed below as a la carte options that can and should be combined in ways that work best for a specific course or group.

Standard 2

Correlating with ACEND's KR1, Standard 2 includes five performance indicators related to locating, retrieving, and managing information. Instruction sessions geared toward lower level nutrition students—freshman or sophomores— should emphasize relevant databases and their content, search strategies, and the importance of organizing and managing information. Lower level nutrition students may be approaching scientific research for the very first time and need basic ideas about where and how to start searching for related literature. Furthermore, lower level students may benefit from explanations about scholarly communication, the peer review process, and the different publication formats that the students may find during their research.

Lower level students, in particular, will appreciate a brief discussion about the information cycle. Emphasizing scholarly journal articles as the main method of communication among scientists and researchers helps students understand why the instruction session may focus more on searching for journal articles and less on searching the library catalog for books. Furthermore, highlighting this distinction between books and journal articles creates the perfect opening for discussing the type of journal articles that students should be looking for: peer-reviewed articles. Often, students experience confusion when they hear journals described as "refereed," "peer-reviewed," and "scholarly," not realizing that these words all essentially mean the same thing. This whole discussion helps students realize that they should be looking for a certain quality and type of information when they start looking for articles for their first research projects.

Additionally, using visual examples of different journal publication formats—loose journal issues, bound periodicals, and electronic journals—helps newer students understand the variety of formats in which scientific literature is published. Once students understand that a journal article might be accessed just in print, in both print and electronic formats, or even as a born-digital article in an electronic journal, they can begin to think about how online indexes and databases can be used to discover and access articles related to their research.

In-class or homework activities can pull together these discussions about scientific research and using relevant databases; one idea for this sort of activity involves asking students to work in small groups to perform basic keyword searches in different databases (see Table 5.3 for examples of databases to use in this activity) in order to find journal articles on general nutrition topics. By assigning a different, relevant database to each small group, librarians can employ what educators call the "jigsaw technique." The jigsaw technique, a teaching method that allows students to become experts in one particular area and then teach it to the rest of the class, works perfectly for a 50 or 75 minute lesson where the librarian must cover several different databases. Students will work in groups to figure out the databases, demonstrate their findings in front of the rest of the class, and then compare their results with the other groups. This sort of activity helps students actively engage with the idea that different databases include different material, and that different databases will be appropriate for different research opportunities. It also ensures that, by the end of the class, students have meaningfully experienced at least one major database in their area of study. Making sure that there is time to follow up these activities with instructor-led discussion is essential so that the instructor can correct any misperceptions, and so that students have the opportunity to ask the instructor any new questions that may have come up during the activity.

Nutrition literature often uses medical terminology, which can intimidate novice researchers as they try to come up with search strategies and interpret their search results. Because of this, information literacy sessions should include discussions about appropriate vocabulary and keywords. Students need to be made aware that authors of scientific articles will very likely be using different, more complicated terms to describe the same research idea that the student has in mind. For example, in PubMed, articles about *heart attacks* will refer to *myocardial infarctions*, while articles about *cancer* may be found by searching with the keyword *neoplasm*. Pointing out the Medical Subject Heading (MeSH) terms on individual PubMed records during class demonstrations can lead into a productive discussion and practice session of brainstorming different synonyms and using Boolean connectors to develop reasonable, inclusive search strings. If students are using the library catalog to search for related books for background information, then pointing out subject headings in the catalog may also help them gain a sense of the terminology being used in a particular area of research. Other resources that will help students brainstorm keywords include medical terminology guides; the Library of Congress Subject Heading "Medicine—Terminology" will locate these types of guides within a library's collection. Quite a few medical terminology educational websites and tutorials also exist on the free web (Table 5.2).

Table 5.2. Resources for Learning About Medical Terminology

Resource	Description	URL
"Understanding Medical Words: A Tutorial from the National Library of Medicine"	This is a brief flash tutorial that includes interactive quizzes.	http://www.nlm.nih.gov/medlineplus/medicalwords.html
Medline Plus's Medical Dictionary	This medical dictionary represents a simple search box for looking up medical terms.	http://www.nlm.nih.gov/medlineplus/mplusdictionary.html
Medical Terminology Course from Des Moines University	This free tutorial walks users through understanding the basics of medical terminology.	http://www.dmu.edu/medterms/
MediLexicon	This website is aimed toward healthcare professionals, offering a dictionary of medical acronyms, abbreviations, and terminology.	http://www.medilexicon.com/
Merck Manual's "Understanding Medical Terms"	This site, associated with Merck, gives a browsable list of common components of medical terms.	http://www.merckmanuals.com/home/about/front/medterms.html

Finally, Standard 2 includes a performance indicator underscoring a student's ability to manage information from various sources. Complex citation management software, like EndNote, may be a bit sophisticated for lower level students, but students often appreciate ideas for simple information management. Recent versions of Microsoft Word include built-in bibliographic management tools; students can use Word to gather citations, insert in-text citations, and create bibliographies with relative ease. Products like Zotero (http://www.zotero.org/), KnightCite (http://www.calvin.edu/library/knightcite/), and BibMe (http://www.bibme.org/) can also help students gain a sense of how to manage references. In-class and homework activities that include students building bibliographies through one of these products can bring together other pieces of Standard 2 and create an effective culminating activity.

Instruction sessions developed for upper level nutrition students—juniors and seniors—should build on the information literacy skills that the students gained earlier in their academic careers. Unfortunately, many students may not have had the opportunity to gain sophisticated information literacy skills by the time they are juniors or seniors. Ideally, librarians would be involved with students throughout the curriculum, from the introductory or orientation classes all the way through the advanced research seminars, helping students evolve their research skills at an appropriate pace. However, it is the disappointing reality that many students receive information literacy instruction only when they are about to embark on an in-depth research project in their research seminars. For this reason, it is important to keep in mind the true level of understanding that students may bring to class. If possible, polling students before a library session to determine their previous experiences with research and library resources can help librarians tailor instruction session to their classes' needs. Most course management systems have polling or quiz functions that can help accomplish this "temperature check", and tools like Google Forms or Poll Everywhere (http://www.polleverywhere.com/) are free and easy to use.

If it turns out that several students have had few prior experiences with research, then emailing or posting a few pre-class tutorials addressing some of the basic ideas that a lower level library instruction class may usually cover can allow the librarian to focus on more advanced information literacy concepts without losing any members of the class. For Standard 2, more advanced instruction sessions should focus on the advanced features of the literature databases and demonstrate more complex citation management systems. For example, advanced database features in PubMed include the search limits, Advanced Search, the "Send to" function, and the Medical Subject Headings (MeSH) database. PubMed provides excellent tutorials and training resources, all of which are available on the PubMed Web site and updated regularly to reflect changes in navigation and search functionality. Demonstrating PubMed's Advanced Search function, which includes the PubMed Search Builder, can help students understand field searching, ultimately reinforcing students' knowledge of how Boolean searching works and how information in databases is organized.

Similarly, demonstrating PubMed's MeSH database can help more advanced students gain a deeper understanding of how information is organized and indexed within PubMed. MeSH is the National Library of Medicine's controlled vocabulary, and uses a hierarchical structure that has the potential to allow students to search at a very specific level. The entire MeSH database of terms is available via PubMed (http://www.nlm.nih.gov/mesh/meshhome. html) and contains a number of helpful tutorials and help sheets. The MeSH database interfaces with the PubMed Search Builder, and allows users to send searches built from MeSH terms directly to PubMed. Researchers will agree that using MeSH to search PubMed is the most efficient way of conducting a literature search; however, MeSH searching is a complex task, and students need to be prepared in order to really benefit from instruction involving MeSH terms and the MeSH database. For this reason, librarians may want to consider addressing MeSH in one-on-one consultations where the librarian can effectively guide a student in his or her use of the MeSH tool and not further confuse the student.

Finally, the "Send to" function in PubMed will help students organize lists of references, and even export files of references to sophisticated citation management software. Instruction sessions might include demonstrations and activities surrounding the use of the "Send to" function to export PubMed references to recommended citation management software, such as Thomson Reuters' End-Note or ProQuest's RefWorks.

Standard 3

Also fitting into ACEND's KR1, Standard 3 focuses on a student's ability to critically evaluate information. Performance indicators include the abilities to evaluate, synthesize, participate in discourse with others, and integrate new information into existing knowledge. As previously discussed, lower level nutrition students often start with a very vague understanding of how scientific literature is produced. In fact, most students are not familiar with the terms "peer-reviewed" and "refereed," essentially failing to realize that "scholarly" articles go through a very specific publication process. Explaining the peer review process in instruction sessions will help students begin to think about the criteria that they can use to evaluate the information that they find. Ideas for in-class activities built around the peer-review process include having students determine whether or not an article has been peer-reviewed, or why it does or does not fit peer review criteria. Most nutrition journals, such as the *Journal of the Academy of Nutrition and Dietetics* (formerly the *Journal of the American Dietetic Association*), have websites with pages dedicated to "Information for Authors" (Journal of the Academy of Nutrition and Dietetics, 2011). On the website for the *Journal of the Academy of Nutrition and Dietetics*, for example, the journal's editorial board lists the specific criteria that peer reviewers use when evaluating an article for publication; distributing this list to students and having them apply the criteria to unidentified articles can facilitate student understanding and valuable discussion.

Activities and discussions surrounding evaluating information can be pushed even further by showing students how to investigate the information resources used in a particular article. Evaluating information in this way is another high-level task that, similar to using MeSH terms, needs to be approached in a cautious manner so that students do not get confused or lost. One way to approach this concept is to start with an article from the popular media and have students work backward to see if they can figure out the scholarly or scientific origins of the article. For example, in September 2010, *Cosmopolitan* magazine posted an article to their website based on a study conducted by Dr. Elizabeth Dennis, Dr. Brenda Davy, and others at Virginia Tech (Ruderman, 2010). The *Cosmopolitan* article reported that subjects lost more weight if they drank two cups of water before each meal; however, the *Cosmopolitan* article failed to report any specifics about the study, such as the very specific age and weight range of the subjects involved in the study. With some guidance, students are able to use PubMed and Web of Science to search with the researcher's name, affiliation, and appropriate keywords in order to locate the original study, which was published in *Obesity*[1]. This "debunking" activity addresses several performance indicators from both Standard 2 and 3: being able to retrieve information, synthesizing main ideas, validating information, and evaluating the procured information (STS-TFILST, 2006). Demonstrating tools like Web of Science's Cited Reference Search and PubMed's author search can add value to these instruction sessions, helping students track entire research trajectories and researchers' various areas of expertise.

At the more advanced level, nutrition students benefit from additional one-on-one sessions. If the instruction session can take place in a computer classroom or if students can bring their own laptops, it may make sense to set aside a large chunk of class time for students to work on their own while the librarian meets individually with each student. If this is not possible during the instruction session, the librarian should stress the potential value of exploring students' research topics in-depth through a one-on-one consultation. This strategy helps students validate and understand the information through "discourse with other individuals... [and] subject-area experts" (STS-TFILST, 2006).

Standard 4

Standard 4 reflects ACEND's KR2; both guidelines focus on students' gaining the ability to communicate effectively and ethically. For lower level nutrition students, instruction sessions should include tutorials and practice sessions revolving around when and how to properly cite information resources. Discussions focusing on the concepts of common knowledge, plagiarism, and copyright will be most helpful for newer students. Additionally, focusing on the mechanics of relevant citation styles will add value to the instruction session; the nutrition discipline most frequently uses the American Psychological Association (APA) style as well as unique styles tied to specific journals in the field. Having students work in groups to create and assess citations according to a particular citation style can supplement and enhance a lecture including citation mechanics; touching on citation management software would also be appropriate at this point.

At the more advanced level, students should have already grasped the mechanics of incorporating and citing information, which means that these instruction sessions should focus on students' ability to effectively communicate information to their target audiences. Recently, the nutrition profession has become a strong player in the social media environment. Blogs, social networking sites, and websites are all communication tools that students will need to be comfortable with as they become emerging dietitians or take on similar roles. Nutrition students need to feel confident in both evaluating and creating social information. Information literacy sessions build this confidence by highlighting these types of information resources; in-class activities can include using regular evaluation guidelines for blog posts or Facebook pages. Furthermore, Standard 4 represents an opportunity for librarians and nutrition faculty to collaborate on assignments that give students the chance to practice writing effective blog or other social media posts that meet positive evaluation criteria (Appendix 5.2). These assignments can also incorporate an element of asking students to include and cite unusual resources, such as images or YouTube videos, or asking students to summarize complex health research for a specific target population.

THE FIRST TASTE: CORE CONCEPTS FOR NON-SCI/TECH LIBRARIANS

Human nutrition programs present all types of librarians with the opportunity to work in interdisciplinary research and instruction environments. Because of the interdisciplinary nature of the nutrition profession, librarians coming from fields such as the humanities, social sciences, and business have unique perspectives and proficiencies that nutrition professionals often find very valuable. For non-sci/tech librarians just starting to work with nutrition faculty and students, successful collaboration remains simply a matter of determining where their special expertise can merge with the disciplinary goals and curricula of nutrition programs. However, many non-sci/tech librarians may need to enter a nutrition classroom or consultation session with faculty before they have time to gain a really strong sense of the subject area and understand the librarian's role within that context.

Librarians entering a nutrition classroom for the first time will need a working knowledge of the core resources for the field of nutrition. Table 5.3 outlines core resources, identifying each database as open access or subscription-based; the URLs for the subscription-based resources below point to product information about the resource, while the URLs for the open access resources link directly to the database content.

Open access options may be most attractive for universities with fewer subscription database options and to recommend to graduating students who will lose access to library resources in the near future. Non-sci/tech librarians will want to spend a few minutes becoming familiar with the searching options, fea-

Table 5.3. Core Resources for Nutrition Research

Database	Access	URL
MEDLINE/PubMed	Open access (not full text)	http://www.ncbi.nlm.nih.gov/pubmed
CABDirect	Subscription-based	http://www.cabdirect.org
Web of Science	Subscription-based	http://www.wokinfo.com/products_tools/multidisciplinary/webofscience/
SciVerse Scopus	Subscription-based	http://www.info.sciverse.com/scopus/
Cochrane Reviews	Subscription-based	http://www2.cochrane.org/reviews/
Academy of Nutrition and Dietetics Evidence Analysis Library	Only available to Academy members	http://www.adaevidencelibrary.com
PubMed Central	Open access (full text)	http://www.ncbi.nlm.nih.gov/pmc/
BioMed Central	Open access (full text)	http://www.biomedcentral.com
Directory of Open Access Journals	Open access (full text)	http://www.doaj.org

tures, and controlled vocabularies of the resources they intend to demonstrate in class or recommend to students. For example, PubMed uses Medical Subject Headings (MeSH) to index literature, while the other databases listed use different organizational schemes. Similarly, some of these databases use very specific journal abbreviations and author naming systems; these issues definitely need to be understood and addressed by librarians working with students.

In addition to gaining familiarity with core databases and other resources, non-sci/tech librarians teaching instruction sessions for nutrition classes need to have a working knowledge of the seminal journals in the field. *Journal Citation Reports (JCR)*, a database allowing users to evaluate and compare the impact and authority of journals using citation data, does have a subject category labeled "Nutrition & Dietetics" enabling users to sort a substantial list of nutrition-related journals by impact factor and other criteria. *JCR* offers a good starting point for non-sci/tech librarians to gain a sense of the important journals in the field; if *JCR* is not available, then browsing MEDLINE journals through the National Library of Medicine (NLM) catalog can represent another good option for locating relevant journal titles. The NLM catalog of journals referenced in the NCBI databases (http://www.ncbi.nlm.nih.gov/nlmcatalog/journals) includes about 120 broad subject terms for journals, one of which is "Nutritional Sciences," and will lead users to a list of journals cataloged under this subject term.

Finally, non-sci/tech librarians who anticipate long-term relationships with a nutrition department or program undoubtedly will benefit from developing a plan to gain subject expertise and stay up to date on trends and advances within the scientific discipline and profession of nutrition. Of course, sci/tech and non-sci/tech librarians alike will be able to advance their instructional impact and departmental collaborations by staying current with the nutrition field. This chapter would be remiss not to mention easy ways to stay current and gain expertise. Leveraging technology so that information is delivered, rather than sought out, may be the best way to save time and receive updates. Table 5.4 identifies specific resources for staying current; technologies that can be used for capturing, delivering, and organizing information from these resources include RSS readers, social media tools, blogs, and listservs. Specifically, tools like Google Reader, the RSS function in Microsoft Outlook, Facebook, Twitter, and LinkedIn will bring in feeds from blogs and dynamic web pages, and connect librarians with organizations and professionals from the field of nutrition.

HUNGRY FOR MORE: FREE BITE-SIZED NUTRITION RESOURCES

This chapter has briefly introduced the scientific endeavors and professional efforts coming out of nutrition departments at colleges and universities across the country. Librarians have much to contribute to these endeavors and efforts if able to stay abreast of changes and growth within the discipline. Busy librarians all have multiple responsibilities and, most likely, multiple departments to keep up with. Staying up to date on trends and research in any one department can be time consuming and difficult to prioritize, especially as demands on a busy librarian's time always seem to be increasing while library budgets may be remaining stagnant or even decreasing. Nevertheless, many free, authoritative resources exist that can provide ongoing support for librarians working with nutrition departments (Table 5.4).

These resources, similar to the rest of this chapter, simply represent a sampling of information about research and resources relating to human nutrition. These resources are intended to support all types of librarians who believe that their special training as information professionals and information literacy experts plays an important role in helping their local nutrition departments or programs reach or exceed their disciplinary and professional goals.

NOTES

1. Dennis, E. Dengo, A. L., Comber, D., Flack, K. D., Savla, J., Davy, K. P., and Davy, B. M. (2010). Water consumption increases weight loss during a hypocaloric diet intervention in middle-aged and older adults. *Obesity* 18, 300–307. doi: 10.1038/oby.2009.235

Table 5.4. Free Resources for Keeping Current with Nutrition Trends and Research

Resource	Facebook	Twitter	Blog (s)/RSS feed(s)
Academy of Nutrition and Dietetics	X	X	X
International Food Information Council (IFIC)	X	X	X
Nutrition.gov			X
USDA Food and Nutrition Service			X
Health & Nutrition on USA.gov			X
Food & Nutrition Information Center			X

REFERENCES

Accreditation Council for Education in Nutrition and Dietetics (2008). *2008 Foundation Knowledge and Competencies—Dietitian Education*. Retrieved from http://www.eatright.org/uploadedFiles/CADE/CADE-General-Content/3-08_RD-FKC_Only.pdf

ALA/ACRL/STS Task Force on Information Literacy for Science and Technology [STS-TFILST]. (2006). Information literacy standards for science and engineering/technology. Retrieved from http://www.ala.org/acrl/standards/infolitscitech

Association of College and Research Libraries. (2000). *Information Literacy Competency Standards for Higher Education*. Retrieved from http://www.ala.org/ala/mgrps/divs/acrl/standards/informationliteracycompetency.cfm

Babbitt, K. R. (1997). Legitimizing nutrition education: The impact of the Great Depression. In S. Stage & V. Vincenti (Eds.), *Rethinking home economics: Women and the history of a profession* (pp. 145–162). Ithaca, NY: Cornell University Press.

Journal of the Academy of Nutrition and Dietetics. (2011). *Author Guidelines*. Retrieved from http://www.ANDjournal.org/authorinfo

Nyhart, L. K. (1997). Home economists in the hospital, 1900–1930. In S. Stage & V. Vincenti (Eds.), *Rethinking home economics: Women and the history of a profession* (pp. 125–144). Ithaca, NY: Cornell University Press.

Ruderman, Z. (2010, September 15). The 60-second trick that will help you shed lbs. *Cosmopolitan.com*. Retrieved from http://www.cosmopolitan.com/celebrity/news/drink-two-glasses-of-water-to-lose-weight-study

Stage, S. (1997). Ellen Richards and the social significance of the home economics movement. In S. Stage & V. Vincenti (Eds.), *Rethinking home economics: Women and the history of a profession* (pp. 17–33). Ithaca, NY: Cornell University Press.

APPENDIX 5.1. ACEND GUIDELINES

COMMISSION ON ACCREDITATION FOR DIETETICS EDUCATION
2008 FOUNDATION KNOWLEDGE AND COMPETENCIES – DIETITIAN EDUCATION

Individuals interested in becoming Registered Dietitians should expect to study a wide variety of topics focusing on food, nutrition and management. These areas are supported by the sciences: biological, physiological, behavioral, social and communication. Becoming a registered dietitian involves a combination of academic preparation, including a minimum of a baccalaureate degree, and a supervised practice component.

The foundation knowledge requirements will be the focus of the academic component of dietitian education, either in a Didactic Program in Dietetics or a Coordinated Program accredited by the Commission on Accreditation for Dietetics Education, the accrediting agency for the American Dietetic Association. These requirements may be met through separate courses, combined into one course, or as part of several courses as determined by the college or university sponsoring a CADE-accredited program.

Competence to practice dietetics is achieved through a CADE-accredited supervised practice component, either in a baccalaureate or masters degree Coordinated Program or a post-baccalaureate Dietetic Internship. Competency statements specify what every dietitian should be able to do at the beginning of his or her practice career. The competency statements build on the foundation knowledge necessary for the entry-level practitioner to perform reliably at the level indicated. A concentration area is added to the basic competencies so that a supervised practice program can prepare graduates for identified market needs. Thus, all entry-level dietitians will have the basic competencies and additional competencies according to the concentration area completed.

FOUNDATION KNOWLEDGE FOR DIDACTIC CURRICULUM CONTENT	COMPETENCIES DIETITIAN SUPERVISED PRACTICE
1: Scientific and Evidence Base of Practice: integration of scientific information and research into practice	
Knowledge Requirement	Competencies/Learning Outcomes
	Upon completion of supervised practice, graduates are able to:
KR 1.1. The curriculum must reflect the scientific basis of the dietetics profession and must include research methodology, interpretation of research literature and integration of research principles into evidence-based practice.	SP 1.1 Select appropriate indicators and measure achievement of clinical, programmatic, quality, productivity, economic or other outcomes
KR 1.1.a. Learning Outcome	SP 1.2 Apply evidence-based guidelines, systematic reviews and scientific literature (such as the ADA Evidence Analysis Library, Cochrane Database of Systematic Reviews and the U.S. Department of Health and Human Services, Agency for Healthcare Research and Quality, National Guideline Clearinghouse Web sites) in the nutrition care process and model and other areas of dietetics practice
Students are able to demonstrate how to locate, interpret, evaluate and use professional literature to make ethical evidence-based practice decisions.	SP 1.3 Justify programs, products, services and care using appropriate evidence or data
KR 1.1.b. Learning Outcome	SP 1.4 Evaluate emerging research for application in dietetics practice
Students are able to use current information technologies to locate and apply evidence-based guidelines and protocols; for example, the ADA Evidence Analysis Library, Cochrane Database of Systematic Reviews and the U.S. Department of Health and Human Services, Agency for Healthcare Research and Quality, National Guideline Clearinghouse Web sites.	SP 1.5 Conduct research projects using appropriate research methods, ethical procedures and statistical analysis

APPENDIX 5.1. ACEND GUIDELINES (CONTINUED)

COMMISSION ON ACCREDITATION FOR DIETETICS EDUCATION
2008 FOUNDATION KNOWLEDGE AND COMPETENCIES – DIETITIAN EDUCATION

FOUNDATION KNOWLEDGE FOR DIDACTIC CURRICULUM CONTENT	COMPETENCIES DIETITIAN SUPERVISED PRACTICE
2: Professional Practice Expectations: beliefs, values, attitudes and behaviors for the professional dietitian level of practice.	
Knowledge Requirements	Competencies/Learning Outcomes
	Upon completion of supervised practice, graduates are able to:
KR 2.1. The curriculum must include opportunities to develop a variety of communication skills sufficient for entry into pre-professional practice.	SP 2.1 Practice in compliance with current federal regulations and state statutes and rules, as applicable and in accordance with accreditation standards and the ADA Scope of Dietetics Practice Framework, Standards of Professional Performance and Code of Ethics for the Profession of Dietetics
KR 2.1.a. Learning Outcome	
Students are able to demonstrate effective and professional oral and written communication and documentation and use of current information technologies when communicating with individuals, groups and the public.	SP 2.2 Demonstrate professional writing skills in preparing professional communications (e.g. research manuscripts, project proposals, education materials, policies and procedures)
KR 2.1.b. Learning Outcome	SP 2.3 Design, implement and evaluate presentations considering life experiences, cultural diversity and educational background of the target audience
Students are able to demonstrate assertiveness, advocacy and negotiation skills appropriate to the situation.	SP 2.4 Use effective education and counseling skills to facilitate behavior change
KR 2.2. The curriculum must provide principles and techniques of effective counseling methods.	SP 2.5 Demonstrate active participation, teamwork and contributions in group settings
KR 2.2.a. Learning Outcome	SP 2.6 Assign appropriate patient care activities to DTRs and/or support personnel considering the needs of the patient/client or situation, the ability of support personnel, jurisdictional law, practice guidelines and policies within the facility
Students are able to demonstrate counseling techniques to facilitate behavior change.	SP 2.7 Refer clients and patients to other professionals and services when needs are beyond individual scope of practice
KR 2.3. The curriculum must include opportunities to understand governance of dietetics practice, such as the ADA Scope of Dietetics Practice Framework, the Standards of Professional Performance and the Code of Ethics for the Profession of Dietetics; and interdisciplinary relationships in various practice settings.	SP 2.8 Demonstrate initiative by proactively developing solutions to problems.
	SP 2.9 Apply leadership principles effectively to achieve desired outcomes
	SP 2.10 Serve in professional and community organizations
KR 2.3.a. Learning Outcome	SP 2.11 Establish collaborative relationships with internal and external stakeholders, including patients, clients, care givers, physicians, nurses and other health professionals, administrative and support personnel to facilitate individual and organizational goals
Students are able to locate, understand and apply established guidelines to a professional practice scenario.	
KR 2.3.b. Learning Outcome	SP 2.12 Demonstrate professional attributes such as advocacy, customer focus, risk taking, critical thinking, flexibility, time management, work
Students are able to identify and describe the roles of others with whom the Registered Dietitian collaborates in the delivery of food and nutrition services.	

APPENDIX 5.1. ACEND GUIDELINES (CONTINUED)

COMMISSION ON ACCREDITATION FOR DIETETICS EDUCATION
2008 FOUNDATION KNOWLEDGE AND COMPETENCIES – DIETITIAN EDUCATION

FOUNDATION KNOWLEDGE FOR DIDACTIC CURRICULUM CONTENT	COMPETENCIES DIETITIAN SUPERVISED PRACTICE
	SP 2.13 Perform self assessment, develop goals and objectives and prepare a draft portfolio for professional development as defined by the Commission on Dietetics Registration
	SP 2.14 Demonstrate assertiveness and negotiation skills while respecting life experiences, cultural diversity and educational background

3: Clinical and Customer Services: development and delivery of information, products and services to individuals, groups and populations

Knowledge Requirements	Competencies/Learning Outcomes
	Upon completion of the SP, graduates are able to:
KR 3.1. The curriculum must reflect the nutrition care process and include the principles and methods of assessment, diagnosis, identification and implementation of interventions and strategies for monitoring and evaluation.	SP 3.1 Perform the Nutrition Care Process (a through d below) and use standardized nutrition language for individuals, groups and populations of differing ages and health status, in a variety of settings
KR 3.1.a. Learning Outcome	SP 3.1.a. Assess the nutritional status of individuals, groups and populations in a variety of settings where nutrition care is or can be delivered
Students are able to use the nutrition care process to make decisions, to identify nutrition-related problems and determine and evaluate nutrition interventions, including medical nutrition therapy, disease prevention and health promotion.	SP 3.1.b. Diagnose nutrition problems and create problem, etiology, signs and symptoms (PES) statements
KR 3.2 The curriculum must include the role of environment, food, nutrition and lifestyle choices in health promotion and disease prevention.	SP 3.1.c. Plan and implement nutrition interventions to include prioritizing the nutrition diagnosis, formulating a nutrition prescription, establishing goals and selecting and managing intervention
KR 3.2.a. Learning Outcome	SP 3.1.d. Monitor and evaluate problems, etiologies, signs, symptoms and the impact of interventions on the nutrition diagnosis
Students are able to apply knowledge of the role of environment, food and lifestyle choices to develop interventions to affect change and enhance wellness in diverse individuals and groups	SP 3.2 Develop and demonstrate effective communications skills using oral, print, visual, electronic and mass media methods for maximizing client education, employee training and marketing
KR 3.3. The curriculum must include education and behavior change theories and techniques.	SP 3.3 Demonstrate and promote responsible use of resources including employees, money, time, water, energy, food and disposable goods.
KR 3.3.a. Learning Outcome	SP 3.4 Develop and deliver products, programs or services that promote consumer health, wellness and lifestyle management merging consumer desire for taste, convenience and economy with nutrition, food safety and health messages and interventions
Students are able to develop an educational session or program/educational strategy for a target population.	

APPENDIX 5.1. ACEND GUIDELINES (CONTINUED)

COMMISSION ON ACCREDITATION FOR DIETETICS EDUCATION
2008 FOUNDATION KNOWLEDGE AND COMPETENCIES – DIETITIAN EDUCATION

FOUNDATION KNOWLEDGE FOR DIDACTIC CURRICULUM CONTENT	COMPETENCIES DIETITIAN SUPERVISED PRACTICE
	SP 3.5 Deliver respectful, science-based answers to consumer questions concerning emerging trends
	SP 3.6 Coordinate procurement, production, distribution and service of goods and services
	SP 3.7 Develop and evaluate recipes, formulas and menus for acceptability and affordability that accommodate the cultural diversity and health needs of various populations, groups and individuals

4: Practice Management and Use of Resources: strategic application of principles of management and systems in the provision of services to individuals and organizations

Knowledge Requirements	Competencies/Learning Outcomes
	Upon completion of the SP, graduates are able to:
KR 4.1.The curriculum must include management and business theories and principles required to deliver programs and services.	SP 4.1 Use organizational processes and tools to manage human resources
KR 4.1.a. Learning Outcome	SP 4.2 Perform management functions related to safety, security and sanitation that affect employees, customers, patients, facilities and food
Students are able to apply management and business theories and principles to the development, marketing and delivery of programs or services.	SP 4.3 Apply systems theory and a process approach to make decisions and maximize outcomes
KR 4.1.b. Learning Outcome	SP 4.4 Participate in public policy activities, including both legislative and regulatory initiatives
Students are able to determine costs of services or operations, prepare a budget and interpret financial data.	SP 4.5 Conduct clinical and customer service quality management activities
KR 4.1.c. Learning Outcome	SP 4.6 Use current informatics technology to develop, store, retrieve and disseminate information and data
Students are able to apply the principles of human resource management to different situations	SP 4.7 Prepare and analyze quality, financial or productivity data and develops a plan for intervention
KR 4.2. The curriculum must include content related to quality management of food and nutrition services.	SP 4.8 Conduct feasibility studies for products, programs or services with consideration of costs and benefits
KR 4.2.a. Learning Outcome	SP 4.9 Obtain and analyze financial data to assess budget controls and
Students are able to apply safety principles related to food, personnel and consumers.	
KR 4.2.b. Learning Outcome	
Students are able to develop outcome measures, use informatics principles and technology to collect and analyze	

APPENDIX 5.1. ACEND GUIDELINES (CONTINUED)

COMMISSION ON ACCREDITATION FOR DIETETICS EDUCATION
2008 FOUNDATION KNOWLEDGE AND COMPETENCIES – DIETITIAN EDUCATION

FOUNDATION KNOWLEDGE FOR DIDACTIC CURRICULUM CONTENT	COMPETENCIES DIETITIAN SUPERVISED PRACTICE
data for assessment and evaluate data to use in decision-making	maximize fiscal outcomes
KR 4.3. The curriculum must include the fundamentals of public policy, including the legislative and regulatory basis of dietetics practice.	SP 4.10 Develop a business plan for a product, program or service including development of a budget, staffing needs, facility requirements, equipment and supplies.
KR 4.3.a. Learning Outcome	SP 4.11 Complete documentation that follows professional guidelines, guidelines required by health care systems and guidelines required by the practice setting.
Students are able to explain the impact of a public policy position on dietetics practice.	SP 4.12 Participate in coding and billing of dietetics/nutrition services to obtain reimbursement for services from public or private insurers
KR 4.4. The curriculum must include content related to health care systems.	
KR 4.4.a. Learning Outcome	
Students are able to explain the impact of health care policy and administration, different health care delivery systems and current reimbursement issues, policies and regulations on food and nutrition services	

5. Support Knowledge: knowledge underlying the requirements specified above.

SK 5.1. The food and food systems foundation of the dietetics profession must be evident in the curriculum. Course content must include the principles of food science and food systems, techniques of food preparation and application to the development, modification and evaluation of recipes, menus and food products acceptable to diverse groups.

SK 5.2. The physical and biological science foundation of the dietetics profession must be evident in the curriculum. Course content must include organic chemistry, biochemistry, physiology, genetics, microbiology, pharmacology, statistics, nutrient metabolism, and nutrition across the lifespan.

SK 5.3. The behavioral and social science foundation of the dietetics profession must be evident in the curriculum. Course content must include concepts of human behavior and diversity, such as psychology, sociology or anthropology

APPENDIX 5.2. BLOGGING NATIONAL NUTRITION MONTH WITH *NOTES FROM NEWMAN* AND EATRIGHT.ORG

Throughout the month of March 2011, *Notes from Newman* (http://hnfelibrarian.blogspot.com/), the HNFE research and resources blog maintained by the HNFE library liaison, Rebecca Miller, will be participating in Blogging National Nutrition Month, hosted by eatright.org. We are using this opportunity to showcase student submissions on this year's topic: Eat Right with Color. Why should you participate in this opportunity? First of all, you can gain extra credit through your submission. Furthermore, participating in this project will give you some experience developing professional blog posts, give you a publication credit, and help you develop a positive digital identity in the online environment (i.e., when potential employers search for your name, they may come across your blog post as a dazzling articulate example of your professional passion).

If you are interested in participating, please submit your post by February 28, 2011. Your post should meet the requirements posted in the grading scale on page 3. The following is an elaboration of these requirements:

Title: Be sure to construct a headline that clearly summarizes or describes the content of your post. Search engines and RSS feed readers will place the post title in results lists, and it is important for it to convey the general idea behind your blog post in addition to being snappy and interesting. Brevity is important here, so try to construct a title that sounds like a newspaper headline—something that will grab people's attention, and is readable in a glance.

Theme: The ADA has selected the theme "Eat Right with Color" for National Nutrition Month 2011. You can interpret this theme any way you'd like! According to eatright.org, NNM campaign focuses on "the importance of making informed food choices and developing sound eating and physical activity habits." To address the theme and goal of National Nutrition Month, be creative in thinking about how you can communicate the theme use color to tell people about eating well. Eatright.org's National Nutrition Month web page gives some good information and examples about working with this theme: http://www.eatright.org/nnm/.

Additionally, think about ways that your post can promote discussion. Ask some thought-provoking questions, or offer some ideas that could potentially generate comments. Remember, blogs are **social** media tools—fostering social interaction on a blog can foster a deeper sense of communication and learning!

Length: Writing blog posts can be tricky: you need to write something that is short enough that people will be able to read and digest it quickly, but long enough that the post communicates your complete thoughts on the topic. Aim for posts that are around 250 words in length.

Content: As with any paper, please check your blog post for spelling and grammatical correctness. When thinking about language and expression, think about the target audience of your blog post: readers of the National Nutrition Month 2011 blog roll: http://www.eatright.org/nnm/. Think about how you can present your information in a way that will reach your target audience of anticipated readers.

Attributions: Just as you would with a research paper or project, it is important that you correctly cite any information or images that come from an outside source. Since you are required to use the APA style for HNFE 3224, use this style when you are attributing any sources for your blog post. Follow the APA guidelines for this.

Images: Images can really improve the way that a blog post communicates its message. You are required to include at least one image with your blog post; please be mindful of copyright restrictions. Just because an image is available on the web does not mean that it is all right for you to simply take and use it. If you find a picture on the web that you like, be sure to use APA guidelines to **cite its source**. Even better, use a Creative Commons Search to find images that image owners and creators have "set free" for others to use. The Creative Commons Search is available: http://search.creativecommons.org/. Remember, though, even if the image is okay for you to use, it's good practice to attribute the source of the image, since you were not the one that created it. Creating your own image is safe, and requires no attribution, so that's also an option!

Tags: Within a blog, posts can be organized with "tags." In *Notes from Newman*, I always tag posts with at least two or three tags that describe the category or content of the post. For example, when I post a tutorial that has to do with a new database, I might tag that post with these tags: **library resources, library news**, and **technology tools**. For the NNM post(s) that you submit to me, please try to come up with two tags that describe your post. When I publish your submissions, I will also tag all posts with **National Nutrition Month 2011** and **Eat Right with Color**. The rest is up to you! Tags can be one word, or more—just be descriptive. Use my blog as an example for coming up with relevant tags.

Around the Web: Online Identity Management

Last fall, I did an "around the web" post that focused on nutrition in the news. I'd like to make this sort of post more of a regular feature on *Notes from Newman*. On these features, I will link to blog posts from other blog authors (i.e., not me!) and other websites that addressed a particularly interesting or relevant topic. I may add a few words of commentary, but for the most part, I'll simply be delivering a list of links that you may want to check out! If you have any suggestions for posts that I can add or blogs that I can follow, just let me know! Here's the first batch of "around the web" blog posts that you may want to check out. The theme?

Managing identity, privacy, and personal branding in the digital environment:

- Managing Your Scholarly Identity, from The Undergraduate Science Librarian (1/12/2011)
- Virginia Tech Policy Safety Tip: Facebook privacy (2/7/2011)
- A Guide to Protecting Your Online Identity, Mashable (4/21/2009)
- Personal Branding 101, Mashable (2/5/2009)

Image from Lifehacker.com

- Establish and Maintain your Online Identity, Lifehacker (5/10/2010)
- Six Steps for Checking your Facebook Privacy, Chronicle of Higher Education (2/7/2011)
- Your Digital Calling Card: About.me, ProfHacker of The Chronicle (11/17/2010)
- Managing Your Online Identity, Forbes.com (6/2/2009)
- Scholarly Reputation Management Online, Michael Habib (SlideShare)

Posted by Rebecca at 12:19 PM

M ▢ ▢ ▤ ▨ +1 Recommend this on Google

Labels: around the web, in the news

Original expression & creativity: Although a blog may feel like a more casual form of communication, writing a blog post is still a very public opportunity for you to display your communication skills. Your blog post needs to be **original** work, and exhibit creativity in its content and style. Remember, plagiarism is a violation of the Virginia Tech Honor Code, and does not contribute to effective communication.

Submission & publication: Although I am asking for everyone to turn in their blog posts by Monday, February 28, I will be spacing out the blog posts throughout the month of March. If I receive 30 submissions, I'll post one a day! If I receive 15 submissions, I'll post one every other day, and so on. Please submit your post to me via email as a Word attachment. If you are including any images, go ahead and insert the .jpg (the preferred file format) into the Word document, and I will place it in the post in the same way you have arranged it within your Word document. When I post your blog post, I will send you an email with the link to the blog post, which will be completely attributed to you, as the sole author.

Questions? Just let me know!

Grading Scale: Submissions that do not meet the minimum following requirements will not be posted on *Notes from Newman* as part of National Nutrition Month. Submissions that merely meet the requirements will receive 10 points, while submissions that exhibit exceptional creativity, originality, and effectiveness will receive 20 points. All acceptable submissions will be posted on *Notes from Newman* during March 2011.

Attributes	Grading Scale	Grading Scale
200–300 words	Acceptable	Acceptable
Two tags	Acceptable	Acceptable
At least one image with appropriate attribution	Acceptable	Acceptable
2 or fewer spelling/grammatical errors	Acceptable	Acceptable
Original work	Acceptable	Acceptable
Creativity	Acceptable	Exceptional
Appropriate for target population	Acceptable	Exceptional
	10 points	20 points

IMPORTANT INFORMATION LITERACY STANDARDS FOR PATENTS

1. The information literate student determines the nature and extent of the information needed.
    3. Has a working knowledge of the literature of the field and how it is produced.
        d. Is knowledgeable of sources that are specific to the field, e.g. manuals, handbooks, patents, standards, material/equipment specifications, current rules and regulations, reference material routinely used in industry, manuals of industrial processes and practices, and product literature.
    4. Considers the costs and benefits of acquiring the needed information.
        d. Recognizes the importance of a variety of information research areas that can be used to gain competitive advantage, track new products, improve processes, and monitor competitors and their marketing strategies. Some examples would be consulting with experts and consultants in a field, research into licensing opportunities, and patent and intellectual property research.

2. The information literate student acquires needed information effectively and efficiently.
    2. Constructs and implements effectively designed search strategies.
        b. Identifies keywords, synonyms and related terms for the information needed and selects an appropriate controlled vocabulary specific to the discipline or information retrieval system.
        c. Uses other methods of search term input such as structure searching and image searching, specific to the discipline or information retrieval system.

3. The information literate student critically evaluates the procured information and its sources, and as a result, decides whether or not to modify the initial query and/or seek additional sources and whether to develop a new research process.
    2. Selects information by articulating and applying criteria for evaluating both the information and its sources.
        g. Recognizes the cultural, physical, or other context within which the information was created, and understands the impact of context on interpreting the information.

IMPORTANT INFORMATION LITERACY STANDARDS FOR PATENTS

CHAPTER 6

6. Determines whether the initial query should be revised.
    c. Reviews information retrieval sources used and expands to include others as needed.

4. The information literate student understands the economic, ethical, legal, and social issues surrounding the use of information and its technologies and either as an individual or as a member of a group, uses information effectively, ethically, and legally to accomplish a specific purpose.
    1. Understands many of the ethical, legal and socio-economic issues surrounding information and information technology.
        d. Demonstrates an understanding of intellectual property, copyright, and fair use of copyrighted material and research data.

5. The information literate student understands that information literacy is an ongoing process and an important component of lifelong learning and recognizes the need to keep current regarding new developments in his or her field.
    1. Recognizes the value of ongoing assimilation and preservation of knowledge in the field.
        a. Recognizes that, for a professional, it is necessary to keep up with new developments that are published in the literature of the field.

INTELLECTUAL PROPERTY: PATENTS

John Meier
Science Librarian
The Pennsylvania State University

INTRODUCTION

In recent years a movement has begun in engineering and moved across all sciences and into other disciplines to incorporate the fundamentals of entrepreneurship into academic curricula. Intellectual property is a vital component of technology in entrepreneurship. In the United States the three main types of intellectual property—copyright, trademarks, patents—were established in Article I Section 8 of the Constitution "to promote the progress of science and useful arts by securing for limited times to authors and inventors the exclusive right to their respective writings and discoveries." Copyright gives authors control over their original, creative works. Trademarks are words, images, or a combination used to identify a product or service currently used in commerce. Patents are a public registration of an invention, design or method that excludes others from making, using or selling that technology (Shackle, 2009).

By this definition a patent can be for a physical object, such as a lawnmower; a process, such as a chemical reaction; or even a design used on an athletic shoe. With so many forms of intellectual property it is easy to confuse copyright, trademarks, and patents. One device that includes all three intellectual properties and multiple patents is the Apple iPod. The "iPod" name and the apple logo image are both examples of trademarks. The music or videos contained on the iPod are protected by copyright. Finally, the iPod involves both utility patents and design patents. The scroll wheel is an example of a utility patent, which only protects functional innovations. Design patents only protect the style and appearance of an invention, such as the iPod's appearance.

Engineers and scientists encounter patents frequently, and therefore students and professors in all technology fields must use them. The need for patent information extends beyond the traditional fields of engineering and chemistry (MacMillan and Shaw, 2008) into all the sciences, including life science (MacMillan and Thuna, 2010), material science and agriculture. This entrepreneurial era requires traditional science and technology information literacy skills and a specific set of literacies that relate to intellectual property information, particularly patent information literacy. Students who already have a high familiarity with basic information literacy will find many concepts portable to patents.

These needs present some great challenges and opportunities for librarians, who have traditionally been the keepers of patent information as US Patent and Trademark Resource Center (PTRC) librarians (USPTO 2011) or as the subject-specialist liaisons for scientists and engineers. Although there have been many improvements in access to these government documents, including international treaties and standards, patents are part of a diverse and complex information landscape that requires specialized skills for finding and using this type of information. Another obstacle to overcome is the traditional subject-oriented relationship between libraries and educators, given that patent information literacy is needed across many science and engineering disciplines. Despite these challenges, librarians have the opportunity to be excellent guides to the patent information landscape.

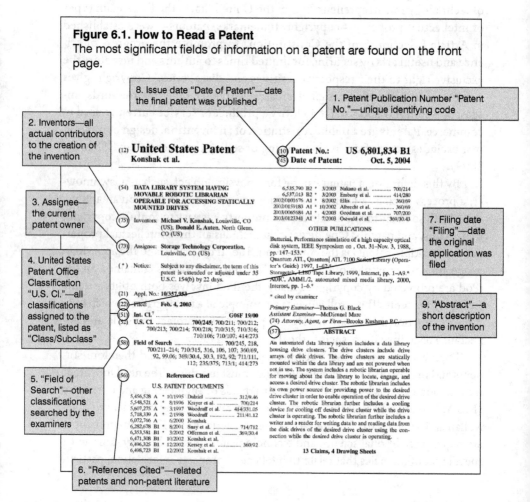

Figure 6.1. How to Read a Patent
The most significant fields of information on a patent are found on the front page.

Figure 6.1. How to Read a Patent (continued)
The remainder of a patent can be a few pages or hundreds of pages and consists of the following.
- Background of the invention describing the technical field and earlier inventions
- Summary of the invention that describes the current proposal
- Detailed Description that includes all information necessary to create or implement the invention including precise data and measurements
- Drawings and descriptions to show the invention
- Claims that are the legal boundaries of protection granted, often patent attorneys are consulted in the writing of patent claims.

SEARCHING FOR PATENTS

The major source of patents is usually the patent office itself, which in the United States is the United States Patent and Trademark Office (USPTO). Historically patents were only published in paper form, but they are now born digital in a number of formats. While they are public documents, patents are also sold to commercial database corporations, which either add them to other documents in an aggregate database or add additional information such as metrics or forecasting. Consequently patents will show up in the search results of many prominent science and technology databases including Compendex, Inspec, Scifinder Scholar, and dozens more. (Baldwin, 2008). Two major sources for United States patent information are the USPTO's Patent Full-Text and Image Database (PatFT) system (2010) and Google Patents (Google, 2011). Many more open access resources are freely available online (LaCourse, 2010) such as Freepatentsonline (http://www.freepatentsonline.com) and Patent Lens (http://www.patentlens.net).

The USPTO system for searching patents provides access to all of their patent records starting in 1790, but has limited indexing before 1976. The database is current with all USPTO internal systems and immediately reflects changes to issued patents. Searching in certain fields, such as date, requires the correct query syntax for a successful search. Results are presented as HTML documents or TIFF images that require a special viewer.

Google Patents is populated by image scans of all USPTO patents since 1790 and all the documents are available freely as PDF downloads. Keyword searching is possible across the entire database, plus Google Patents provides an advanced search mode that allows field searching. The scanning process sometimes results in errors due to poor optical character recognition (i.e., 1880 instead of 1980). In addition, patents and applications may be delayed up to a few months before they are entered into the Google system.

When searching for patents, the simplest and most common strategy is the "known item" search for a single patent, most often by patent number. This strategy is usually used in following a citation or reference to a patent. Field searching is generally effective in a "known item" search if additional data such as inventor name is known. The only exception to this is searching by title, which is generally the most difficult field by which to search. For example, the original patent for Teflon is titled "Tetrafluoroethylene polymers" and the brand name is listed nowhere in the patent. The most effective way to search patents is by classification searching (see "7 Step Strategy for Searching by Classification") as it has both a high precision (only relevant results returned) and high recall (all desired results returned). The patent's classifications (similar to subject headings) are the technology areas to which a patent is assigned, and they are strictly controlled by patent offices. Inventors who need to be sure no previous patent duplicates their innovations or researchers needing a thorough set of results often use classification searching. Keyword searching can be ineffective in patent information seeking, due to the codes and abbreviations in the documents as well as the broad and technical language used in the descriptions.

Figure 6.2. Seven Step Strategy for Searching by Classification

The seven step strategy for classification searching is recommended by the USPTO.

1. Brainstorm keywords related to the purpose, use and composition of the invention.
2. Look up the words in the Index to the U.S. Patent Classification to find potential class/subclasses (USPTO, 2011b).
3. Verify the relevancy of the class/subclasses by using the Classification Schedule in the Manual of Classification (USPTO, 2011c).
4. Read the Classification Definitions to verify the scope of the subclasses and note "see also" references.
5. Search the Issued Patents and the Published Applications databases by "Current US Classification" and access full-text patents and published applications.
6. Review the claims, specifications and drawings of documents retrieved for relevancy.
7. Check all references and note the "U.S. Cl." and "Field of Search" areas for additional class/subclasses to search.

Accessed at http://www.uspto.gov/products/library/ptdl/services/step7.jsp

---

APPLYING THE STANDARDS

It is important to advocate for the literacy needs of students, and the Information Literacy Standards for Science/Engineering and Technology (ACRL, 2006) provide an excellent framework for discussion. The following guidelines address the range of skills needed by entrepreneurs, a group which will include students, researchers and student-inventors (Table 6.1).

Table 6.1. Appropriate Standards, Performance Indicators and Outcomes from the Standards (STS-TFILST, 2006) with Suggested Activities

Activity	Standard/ Performance Indicator/Outcome
In-class discussion of science and technology research and the different methods of dissemination	1.3.d
Explanation of costs and benefits to the patent process including licensing	1.4.d
Printed patent handout with explanation of patent classification schemes	2.2.b
Demonstration with chemistry databases with structure searching tools	2.2.c
Examples of the patent application process and inventor information	3.2.g
In-class activity, comparison of results from USPTO and Google Patents	3.6.c
Examples in lecture followed by in-class quiz or post-class assignment	4.1.d
Demonstration of current awareness tools and examples of competitive intelligence in industry	5.1.a

Identifying the Need for Patent Information

Information need is critical to evaluate as patents can be used by researchers to discover the workings of a technology and by inventors needing to secure legal defense of their ideas. A discussion about the need for incorporating patent information into a course should look at a number of factors including learning outcomes, the current level of student literacy in all aspects of intellectual property, and unique characteristics of the technology field. In an introductory course, the learning outcomes likely will focus only on an awareness of different intellectual property types including patents. Another course may focus specifically on patents as they are involved in a project or paper in the class. Even patents as literature, a traditional research source, could be the rationale for teaching patent information literacy to students.

Each field of science or engineering requires unique patent information, which should inform the focus of teaching and learning. An engineering design course may be interested in patent drawings alone and the process involved in producing them. A chemistry course would likely ask students to seek patent information using chemical compound identifiers or structures. Patents may be freely available, public information, but additional specialized data such as chemical identifiers, corporate information, patent metrics, or translations of foreign patents come at a significant cost.

Ethics and Intellectual Property

In few other areas are the ethical and legal uses of information more vital than in intellectual property, where the ultimate pursuit may be exclusive legal rights. In the United States each of the three main types of intellectual property—copyright, trademarks, patents—have different requirements, legal protections, and durations. Confusion about patents, copyright and trademarks is not uncommon because the boundaries between them are often unclear to the uninitiated. For example, copyright currently lasts for the life of the author plus 70 years, while trademarks can be renewed indefinitely. Patents are limited to 20 years after the filing date in order to place technological innovation into the public domain more quickly. Familiarity with all three types of intellectual property should be a fundamental for all librarians, who should use caution when addressing certain intellectual property questions as it may be considered legal advice and require professional credentials. More examples and explanations of intellectual property and information about how to help library users locate information can be found in the book *Intellectual Property: Everything the Digital-age Librarian Needs to Know* by Timothy Wherry (2008).

Keeping Up with Innovation

Finally, lifelong learning and application to the practicing professional is the core value of the academic entrepreneurship movement. The ability to seek the information found in patents is valuable to business, legal, science and engineering professionals. Because the information sources discussed in this chapter are freely available on the web, it is possible for students to take full advantage of them later in their careers.

Unique current awareness services are available to benefit those seeking patent information. Google Patents (Google, 2011) provides subscriptions to RSS (Really Simple Syndication) feeds based on search queries. When new patents or applications that satisfy the search are added to the system they are sent to subscribers. Free Patents Online (2011) is one example of other websites that provide RSS feeds and email alerts by technology area or classification. This site makes very current applications and issued patents available in PDF format. While RSS feeds are available to anyone visiting the site, a free account registration is often required for other functions such as email alerts and batch downloading of patents.

INCORPORATING PATENT INFORMATION LITERACY INTO THE CURRICULUM: EXAMPLES, ACTIVE LEARNING TECHNIQUES AND ASSIGNMENTS

As an introduction to patent literature, it is best to begin with intellectual property as a whole. One great example for discussion of the differences between copyright, trademarks, and patents is the Apple iPod as discussed previously. An image of this single device can be used to solicit ideas from a classroom about which parts and features are protected by which intellectual property right. After giving definitions for copyright, trademarks, design patents, and utility patents, ask students to identify the elements in the iPod product.

Software patents are another great example to use when discussing the differences between copyright, trademarks, and patents. If a programmer were to write a computer program, he or she could be protected by each of these intellectual property rights. First, the code as written is protected by copyright, ensuring that it cannot be duplicated or copied directly. Secondly, the software name can be trademarked, so that no competing products can falsely claim to be the same. Finally, the function of the program, if patented, will restrict others from writing their own code to do the same functions, a protection even against reverse engineering.

Building on a basic discussion of intellectual property, it is important to tailor the remainder of the content to any course assignments or specific course learning outcomes (for example, an engineering entrepreneurship course may need to assess the students' ability to perform classification searching). As mentioned previously this is a vital skill to patent information retrieval, so it is advantageous to have an example assignment ready or to develop one in partnership with a course instructor. Classification searching can be taught by demonstration, but in order to increase teaching effectiveness, a hands-on exercise should also be attempted. Small groups of students can be directed to use different starting points for a classification search and asked to report back success or have a competitive race to find an example patent.

One example of integrated patent information literacy that has been a great success involves a one-credit course on the "Chemical Literature" at Pennsylvania State University taught by the chemistry liaison librarian. There are two one-hour class periods devoted to patents and intellectual property along with suggested readings and a graded assignment. The first class session is a basic discussion of intellectual property as described above along with demonstrated examples of known item and classification searches. The assignment (see Appendix 6.1) is distributed at the end of the first class and collected at the beginning of the next week's class.

The second class session begins with a review of the homework assignment, but the majority of the time is devoted to demonstrated searches for patents in the chemical databases SciFinder Scholar and Reaxys. These are emphasized over the USPTO and Google Patent databases because the patents included in these resources are limited to those related to chemistry dating back to 1828 for SciFinder Scholar and 1920 for Reaxys, and they include additional indexing and data. Both databases allow users to search for chemicals using name, formula, and even chemical structure. Both also allow patent searches for "prophetic" chemicals, which are newly identified substances that have been described but not yet produced. Because there is no requirement that experimental evidence or working models of an invention be included in a patent application, prophetic patents can be granted for such chemicals that have yet to be tested.

INTEGRATING PATENT INFORMATION LITERACY ACROSS DISCIPLINES

Patents are a multidisciplinary subject, which can be challenging for librarians working in a traditional subject-specialist model where they are the lone contact between the library and a specific set of academic departments. If possible, a suitable solution is to identify some librarians as format specialists for types of material such as newspapers, government documents and patents. The expertise of these librarians should be highlighted for both library users and staff so that requests reach them through referral.

Other proactive ways to reach broadly across the curriculum should be explored. For example, whenever a patent information literacy session or event is scheduled, it can be advertised to the entire campus and perhaps the local community as well. Rather than focusing on a department or subject library, consider using the main library's news stream. In addition, identification of existing student and campus organizations that focus on entrepreneurship can be used to approach instructors and advertise the library's expertise in this area.

Finally, individual contact with instructors across the campus is the most effective method for identifying the greatest needs for patent information literacy in the curriculum. Investigating syllabi, descriptions of courses and requirements for minors or majors is an effective way to survey the landscape. Attending entrepreneurship and intellectual property events across campus creates opportunities for networking and face-to-face contact with key administrators and instructors. A librarian can serve as a key facilitator in bringing together disparate parties into a common conversation about intellectual property and information literacy.

CONCLUSION

This chapter presents an introduction to patent information literacy for librarians and includes examples of integrating it into library instruction. However, the

Figure 6.3. Additional Resources

Arizona State University LibGuides—Patents
http://libguides.asu.edu/patents

Pennsylvania State University Libraries Patent Search Tutorial
http://www.libraries.psu.edu/psul/pams/patent.html

USPTO Public Training Portal
http://www.USPTOTraining.gov

particular needs of students in any institution need be addressed locally in order to meet the specific information literacy goals of the curriculum and determine how patents fit best into this framework. The USPTO provides help for inventors, including new online courses (see Additional Resources), which address more advanced skills such as accessing patent application information and international patents that are beyond the scope of this chapter. The importance of current awareness and instruction for students is highlighted by the fact that more than 8 million patents exist in the United States alone. This number will continue to increase, and patents will become increasingly complex as new laws, such as the Leahy-Smith America Invents Act, are passed.

REFERENCES

ALA/ACRL/STS Task Force on Information Literacy for Science and Technology [STS-TFILST]. (2006). *Information literacy standards for science and engineering/technology.* Retrieved from http://www.ala.org/acrl/standards/infolitscitech

Baldwin, V. (2008) Patent information in science, technical, and medical library instruction. *Science & Technology Libraries, 28,* 263–270.

Free Patents Online. (2011). *Patent searching and inventing resources.* Retrieved from http://www.freepatentsonline.com/

Google. (2011). Google patents. Retrieved from http://www.google.com/patents/

LaCourse, P. (2010). End-user patent searching using open access sources. *Issues in Science and Technology Librarianship, 60.* Retrieved from http://www.istl.org/10-winter/internet.html

MacMillan, D., & Thuna, M. (2010). Patents under the microscope. *Reference Services Review, 38,* 417–430.

MacMillan, M. & Shaw, L. (2008). Teaching chemistry students how to use patent databases and glean patent information. *Journal of Chemical Education, 85,* 997–999.

Shackle, L. (2009). A short course on patent reference for science and technology librarians. *Issues in Science and Technology Librarianship, 59.* Retrieved from http://www.istl.org/09-spring/experts2.html

United States Patent and Trademark Office [USPTO]. (2010). PatFT. Retrieved from http://patft.uspto.gov/

USPTO. (2011a). Patent and Trademark Resource Center Program. Retrieved from http://www.uspto.gov/products/library/ptdl/index.jsp

USPTO.. (2011b). Index to the USPC. Retrieved from http://www.uspto.gov/web/patents/classification/uspcindex/indextouspc.htm

USPTO. (2011c). Patent Classification Homepage. Retrieved April 7, 2011, from http://www.uspto.gov/web/patents/classification/

Wherry, T. L. (2008). *Intellectual property: Everything the digital-age librarian needs to know.* Chicago: American Library Association.

APPENDIX 6.1. PATENT ASSIGNMENT FOR A CHEMICAL LITERATURE CLASS

1. What is the title of U.S. patent number 2230654? List a common name for this invention.
2. What is the current U.S. classification number for CLASS "nanotechnology"?
3. What are the current U.S. CLASS and SUBCLASS of **Design** patents for microscopes? Pick a patent issued in this area and find the name of the company that is the assignee for it (list their name and the patent number).
4. Find a patent with the U.S. classification **422/51** Calorimeter that expired in 2010 and is now in the public domain (anyone can use it). List the patent number and filing date of patent.
5. You have heard about a recent invention by a Penn State faculty member, Thomas Mallouk, for a microstructure. Find the patent number and the assignee.

IMPORTANT INFORMATION LITERACY STANDARDS FOR INTERDISCIPLINARY SCIENCE COURSES

1. The information literate student determines the nature and extent of the information needed.
   1. Defines and articulates the need for information.
      a. Identifies and/or paraphrases a research topic, or other information need such as that resulting from an assigned lab exercise or project.
   2. Identifies a variety of types and formats of potential sources for information.
      a. Identifies the purpose and audience of potential resources (e.g. popular vs. scholarly, current vs. historical, external vs. internal, primary vs. secondary vs. tertiary).
   3. Has a working knowledge of the literature of the field and how it is produced.
      a. Knows how scientific, technical, and related information is formally and informally produced, organized, and disseminated.
      e. Recognizes that knowledge can be organized into disciplines and combinations of disciplines (multidisciplinary) that influence the way information is accessed and considers the possibility that the literature of other disciplines may be relevant to the information need.

2. The information literate student acquires needed information effectively and efficiently.
   1. Selects the most appropriate investigative methods or information retrieval systems for accessing the needed information.
      b. Investigates the scope, content, and organization of information retrieval systems.
   2. Constructs and implements effectively designed search strategies.
      b. Identifies keywords, synonyms and related terms for the information needed and selects an appropriate controlled vocabulary specific to the discipline or information retrieval system.
      d. Constructs a search strategy using appropriate commands for the information retrieval system selected (e.g., Boolean operators, truncation, and proximity for search engines; internal organizers such as indexes for books)

IMPORTANT INFORMATION LITERACY STANDARDS FOR INTERDISCIPLINARY SCIENCE COURSES

    e.   Implements the search strategy in various information retrieval systems using different user interfaces and search engines, with different command languages, protocols, and search parameters, while recognizing similar search features across the systems (such as: e-mail alerts and save search options, search fields, and controlled vocabulary.)

    f.   Follows citations and cited references to identify additional, pertinent articles.

5.   Extracts, records, transfers, and manages the information and its sources.

    d.   Records all pertinent citation information for future reference by downloading, printing, emailing, or manual notation.   Uses various technologies to manage the information selected and organized, e.g., bibliographic management software.

3.   The information literate student critically evaluates the procured information and its sources, and as a result, decides whether or not to modify the initial query and/or seek additional sources and whether to develop a new research process.

   1.   Summarizes the main ideas to be extracted from the information gathered.

      a.   Applies an understanding of the structure of a scientific paper and uses sections, such as the abstract or conclusion, to summarize the main ideas.

      b.   Selects main ideas from the text.

   2.   Selects information by articulating and applying criteria for evaluating both the information and its sources.

      a.   Distinguishes between primary, secondary, and tertiary sources, and recognizes how location of the information source in the cycle of scientific information relates to the credibility of the information.

   4.   Compares new knowledge with prior knowledge to determine the value added, contradictions, or other unique characteristics of the information.

      c.   Draws conclusions based upon information gathered.

      f.   Integrates new information with previous information or knowledge.

IMPORTANT INFORMATION LITERACY STANDARDS FOR INTERDISCIPLINARY SCIENCE COURSES

g. Determines whether information provides evidence relevant to the information need or research question.

6. Determines whether the initial query should be revised.
    b. Reviews search strategy and incorporates additional concepts as necessary.
7. Evaluates the procured information and the entire process.
    a. Reviews and assesses the procured information and determines possible improvements in the information seeking process.
    b. Applies the improvements to subsequent projects.

4. The information literate student understands the economic, ethical, legal, and social issues surrounding the use of information and its technologies and either as an individual or as a member of a group, uses information effectively, ethically, and legally to accomplish a specific purpose.
    2. Follows laws, regulations, institutional policies, and etiquette related to the access and use of information resources.
        f. Demonstrates an understanding of what constitutes plagiarism and does not represent work attributable to others as his/her own. This includes the work of other members of research teams.
    3. Acknowledges the use of information sources in communicating the product or performance.
        a. Selects an appropriate documentation style for each research project and uses it consistently to cite sources.

INTERDISCIPLINARY SCIENCE COURSES: REMOTE SENSING

Linda Blake
Science Librarian and Electronic Journals Coordinator
West Virginia University

Timothy A. Warner
Professor of Geography and Geology
West Virginia University

INTRODUCTION

Increasingly, research in science, technology, engineering, and mathematics (STEM) does not fall within a distinct discipline, but instead is interdisciplinary in nature. Chemists studying carbohydrates and their effects on cancer, engineers developing devices that use biological information to establish identity, and geologists using satellite imagery to locate new mineral deposits, all work across multiple disciplines. Librarians seeking to teach information literacy to students in such cross-cutting and multidisciplinary scientific fields face particular challenges. First, the information is often archived in a wide array of discipline-specific resources, and the students need skills in database searching, particularly the ability to focus queries effectively. Second, the interdisciplinary sciences, as with science in general, have a complex hierarchy of reliability of sources, including the popular press, trade literature, gray literature, and various levels of peer-review. To confuse the issue even more, some journals, such as *Nature*, may include mixtures of research and news or opinion pieces. Thus, evaluating the reliability of scientific information based on its context requires a sophisticated understanding of the research milieu, which is typically discipline-specific. A third challenge is that plagiarism issues are complex, since the conventions taught to students in humanities-oriented writing classes may not translate to scientific writing. For example, scientific writing only rarely draws on quotations, despite the challenge of summarizing very precise technical concepts. Identifying general knowledge that does not need to be cited is inherently subjective and based on the audience's subject expertise, resulting in an additional challenge for students. Perhaps most troubling, students may be over-confident in their ability to identify and deal with plagiarism issues (Blake & Warner, 2011). A final challenge for librarians seeking to teach information literacy is that students in the sciences often need to create and manage large and complex bibliographies, suggesting that citation management software may be an important tool for them.

This chapter discusses these aspects of information literacy in the context of remote sensing, an interdisciplinary scientific field. The focus of the chapter is on concepts as well as suggestions and strategies for the non-science librarian with instructional responsibilities in the sciences.

Remote sensing can be defined as the art and science of gaining information about objects using sensors not in contact with the objects (Lillesand, Kiefer, & Chipman, 2008). Examples of remotely-sensed data include the satellite imagery commonly shown in weather forecasts and the images displayed in Google Maps. Remote sensing analysts use such data in a variety of applications, including mapping urban sprawl, locating forest pathogens, and searching for new mineral deposits. The remote sensing scientific literature therefore includes engineering, physics, computer science, geography and an extensive variety of applied disciplines, such as agriculture, biology, geology, and meteorology. Information literacy is important in this field because, as with most technology-oriented disciplines, students need to be prepared for careers in which it will be essential to have the ability to stay current and the skills to research new and emerging topics. Thus, information literacy is not just important for scholarly research, but is rather a basic professional skill that students need prior to entering the workforce.

The authors' experience in information literacy education in remote sensing developed from collaboration over an extended period in teaching information literacy to students in a cross-listed geology and geography class, Introduction to Remote Sensing (Geol/Geog 455) at West Virginia University (WVU). The class enrollment, typically about 35 students, has been diverse, including seniors and both master's and doctoral graduate students, with majors that include geology, geography, environmental geosciences, biology, wildlife management, engineering, and even the humanities. For a number of years, the librarian gave the class a single class/laboratory experience in literature search methods. Starting in 2009, under the auspices of a grant provided by the WVU Libraries (Wilkinson, 2011), the authors incorporated information literacy as a key learning outcome for the class (Blake & Warner, 2011), and this chapter draws extensively upon that experience (See Assignment in Appendix 7.1).

KEY STANDARDS FOR REMOTE SENSING INFORMATION LITERACY

Information Literacy Standards for Science and Engineering/Technology (hereafter, Standards) provide a comprehensive catalog of the information literacy skills modern scientists need (ALA/ACRL/STS Task Force on Information Literacy for Science and Technology [STS-TFILST], 2006). In the context of developing a curriculum within which information literacy is just one component, it is useful to identify key standards to provide a central focus. Table 7.1 lists the key standards, performance indicators and outcomes for an introductory course in remote sensing, based on the experience of working with seniors and graduate students at WVU. These standards were used to shape the content of the lesson plans, teaching strategies, and assessments, as discussed below.

Table 7.1. Summary of Key Standards, Performance Indicators, Outcomes (STS-TFILST, 2006) Together with Associated Class Activities

Standard	Performance Indicator	Outcome	Activity / Leader	Week / Lesson*
1. Determines the nature and extent of the information needed	1. Defines and articulates the need for information	a. Identifies a research topic	In-class exercise / Librarian	1 / 1
	2. Identifies a variety of types of potential sources	a. Identifies the purpose and audience of potential resources	In-class exercise / Librarian	3 / 4
	3. Has a working knowledge of the literature of the field and how it is produced	a. Knows how scientific information is formally and informally produced and disseminated	Lecture / Instructor	2 / 2
		e. Recognizes that knowledge can be organized into disciplines and combinations of disciplines	Lecture and in-class exercises (weeks 2 & 3) / Librarian & Instructor	2 & 3 / 2–4
2. Acquires needed information effectively and efficiently	1. Selects the most appropriate investigative methods	b. Investigates the scope, content and organization of information retrieval systems	In-class exercise / Librarian	3 & 4 / 3–4
	2. Constructs and implements effectively designed search strategies	b. Identifies keywords and synonyms	In-class exercise / Librarian	1, 3, & 4 / 1, 3–4

Standard	Performance Indicator	Outcome	Activity / Leader	Week / Lesson*
Table 7.1. Summary of Key Standards, Performance Indicators, Outcomes (STS-TFILST, 2006) Together with Associated Class Activities				
		d. Constructs a search strategy using appropriate commands	In-class exercise / Librarian	1, 3 & 4 / 1,3–4
		e. Implements the search strategy in various information retrieval systems	Homework exercise / Librarian & Instructor	3–7 / 4
		f. Follows citations and cited references to identify additional, pertinent articles	In-class exercise; Term paper / Librarian & Instructor	3 & 15 / 4
	5. Manages the information and its sources	d. Records all pertinent citation information, for example using bibliographic management software	In-class exercise / Librarian	3 / 4
3. Evaluates the procured information and its sources	1. Summarizes the main ideas to be extracted from the information gathered	a. Applies an understanding of the structure of a scientific paper	Paper reviews, term paper / Instructor	3–6, 15
		b. Selects main ideas from the text	Paper reviews, term paper /Instructor	3–7, 15
	2. Select information, cognizant of source	a. Distinguishes between primary, secondary, and tertiary sources	Paper reviews, term paper / Instructor	3–7, 15
	4. Compares new knowledge with prior knowledge	c. Draws conclusions upon information gathered	Paper reviews, term paper / Instructor	3–7, 15

Standard	Performance Indicator	Outcome	Activity / Leader	Week / Lesson*
		f. Integrates new information with previous information	Paper reviews, term paper / Instructor	3–7, 15
		g. Determines whether information provides evidence relevant to the research question	Term paper / Instructor	15
	6. Determines whether the initial query should be revised	b. Reviews research strategy and incorporates additional concepts as necessary	Term paper / Instructor	15
	7. Evaluates the procured information and the entire process	a. Reviews and determines possible improvements in the information seeking process	Paper reviews, term paper / Instructor	3–7, 15
		b. Applies the improvements to subsequent projects	Paper reviews, term paper / Instructor	3–7, 15
4. Used information effectively, ethically, and legally	2. Follows laws, institutional polices related to the access and use of information resources	f. Demonstrates an understanding of plagiarism	Paper reviews, term paper / Instructor	3–7, 15
	3. Acknowledges the use of information sources	a. Consistently uses an appropriate style to cite resources	Term paper / Instructor	15

Table 7.1. Summary of Key Standards, Performance Indicators, Outcomes (STS-TFILST, 2006) Together with Associated Class Activities

* Lesson numbers only listed for the lessons described in the Appendices.

TEACHING STRATEGIES

After identifying and articulating the information literacy needs of students in the remote sensing class, the authors created four lesson plans with goals, objectives, and teaching strategies (see Appendices 1 through 4) to meet the specific information literacy needs. The class schedule of activities, divided between the course instructor and the librarian, is outlined in Table 7.2. The librarian also created a LibGuide (Blake, 2010) to provide a starting point for accessing all of the sources taught in class.

Table 7.2. Information Literacy Collaboration Schedule for Introduction to Remote Sensing, Geography/Geology 455

Week	Appendix	Instructor	Location	Instruction
1	1	Faculty and Librarian	Classroom	Developing a topic and search statement; academic dishonesty
	2	Faculty	Classroom	Understanding the process of scholarly information production and the structure of a typical scholarly work in the geosciences
2	3	Librarian	Library Lab	Searching subject-area databases effectively; managing citations using a reference management tool
3	4	Librarian	Library Lab	Searching subject-area databases continued; citation searching; and using the reference management tool for in-text citing while writing a paper

Lesson One

The librarian met the students in their regular classroom for the first one-hour lesson. This lesson helped students define their topics both as a research question and as a search statement expressed in keywords and phrases, Boolean operators, and truncation symbols. The librarian demonstrated how to narrow a very broad topic, *remote sensing of forests*; to a restricted topic, *deforestation*; to a narrow topic, *remote sensing of deforestation in tropical regions*. Subsequently, she indicated a good research question based upon the topic: "What trends does remote sensing show for deforestation of the Amazon?" Then, students were given a form and were asked to develop their own research questions with the help of the instructor, the librarian, and the class graduate assistant.

Next, using the research question composed previously, the librarian showed the students how to pick keywords and concepts to use to compose a search statement for searching the databases. She presented two alternative search statements:

- *"remote sensing" and deforestation and Amazon**
- *"remote sensing" and forest* and Amazon**

The librarian used the remainder of the time to discuss academic dishonesty and the role of correctly citing sources in avoiding plagiarism. For an active learning experience, the librarian showed slides of possible scenarios and asked the students to indicate whether they agreed or disagreed. An example of a scenario was:

> To Cite or Not to Cite? While reading an introduction to remote sensing in an encyclopedia, a student learns that humans have always used remote sensing in some form including early examples such as climbing a tree, standing on a hill, or sniffing the air. If used in a paper, should this information be cited?

The students became quite engaged with this activity, and a post-course assessment of their understanding of plagiarism and academic honesty showed that they questioned their previous assumptions about these concepts (Blake & Warner, 2011).

Lesson Two
For the next information literacy session, the course instructor presented a one-hour lecture on information in the context of the geosciences and interdisciplinary science in general. The purpose of this class was to give the students the background to be able to recognize and differentiate the reliability of different types of sources of scientific information. For example, the difference between conference proceedings and journal articles can be subtle for students to understand, but an explanation of the peer review process and how it is typically used in journals versus conference proceedings can provide the necessary context.

To conserve class time, the students were given printed instructions and asked to create EndNote Web and interlibrary loan accounts as homework. They were also asked to install the Capture Firefox plug-in to use for adding web page references to EndNote Web, and to install the Cite While You Write plug-in for Microsoft Word.

Lesson Three
The third session, a one-hour guided, hands-on exercise, was held in the library's computer laboratory. The librarian introduced the students to the WVU Librar-

ies' web page and pointed out some pertinent services such as the help page. She showed them how to access the class LibGuide, which was used as a starting point for the instruction.

Following on the content of lesson two, the course instructor's lecture, the librarian showed the students a slide defining the difference between popular and scholarly resources. Then she projected the sample topic from the first session: *"What trends does remote sensing show for deforestation of the Amazon?"* The students were asked to identify the keywords in the topic. The librarian used these keywords to formulate the search statement to be used in database searching. Additionally, she spent a few minutes explaining that for an interdisciplinary field such as remote sensing, it is necessary to search a number of databases, including those encompassing many disciplines as well as those that are discipline-specific to the content needed by the researcher.

One more task needed to be completed before searching the databases. After logging in to their EndNote Web accounts, students created a group in EndNote Web called *Remote Sensing Practice* in preparation for adding references. Two relevant databases in the EbscoHost platform were selected because of the ease of importing references into EndNote Web. First the students searched Academic Search Complete, which was selected because of its multidisciplinary focus, reflecting the nature of the course content. Then they searched Agricola, an agriculture database, which was selected because it includes articles on deforestation, our sample topic. While searching these databases, students imported references into EndNote Web and moved them to the *Remote Sensing Practice* group.

With the librarian, the students searched additional databases with content particularly relevant to remote sensing topics. The American Geological Institute's GeoRef indexes the geoscience literature found in a number of formats including journal articles, books, and reports. Compendex and Inspec contain references for journal articles, web sites, proceedings, patents, and product information. The librarian demonstrated how to search GeoRef using the search statement:

- remote sensing [in abstract] AND (deforestation OR forest*) [in abstract] AND Amazon* [in abstract].

In this way, the students learned how to narrow their results by searching in the abstract only and by using the *AND* Boolean operator. They learned how to expand their results by using the *OR* Boolean operator and the use of the asterisk as a truncation symbol. For GeoRef, Compendex and Inspec, the students learned how to add references to EndNote Web by saving them to a file and then importing them.

Lesson Four

During the final one-hour session, also held in the library's computer laboratory, students were introduced to citation searching using Web of Science and were shown how to use the references in their EndNote Web accounts to incorporate in-text citing while writing a paper. The librarian started the session by defining the extent and limitations of Web of Science, which provides article citation counts as well as topic indexing. This database indexes the top journals in science, social sciences, and humanities, and the WVU Libraries' subscription only goes back to 1976. She then demonstrated a search using the search statement:

- *"remote sensing" AND (Amazon or Amazonia) AND deforestation.*

The librarian showed the students the various ways results can be refined including by subject area, document type, authors, publication years, and authors' institutions. For example, graduate students interested in finding funding agencies specific to their research could use the Funding Agency sort to identify funding sources by topic.

The second part of teaching Web of Science dealt with how to use citations to track the impact of an article. The librarian showed the students a visual representation of the impact of a key article forward in time by illustrating how the article was cited, and back in time by listing the article's citations. Then, the students were asked to do a search for a key article chosen by the instructor to examine the articles cited in the key article, as well as the articles citing the key article. Choosing a key article that was written by a prominent scholar at the students' institution could be an effective way to engage the students in this exercise.

Instruction for this session was concluded by showing students how to use the EndNote Web plug-in, Cite While You Write, to build a bibliography and create in-text citations while writing a paper in Microsoft Word. While EndNote's Cite While You Write only works with Microsoft Word, there are two versions of the plug-in: one for Microsoft Windows and one for Macintosh operating systems. The students were impressed with this tool. This provided a great culmination of the learning experience – from the beginning, when they learned to pick a topic, and to the end, when they learned to assimilate their discoveries into a paper.

Finally, to practice using what they had learned, students were asked to use their own research questions and research statements, which they had created in the first session, to search a database of their choosing. They imported the resulting references to EndNote Web.

CREATING/CEMENTING LIBRARIAN-FACULTY RELATIONSHIPS

An information literacy program can only be successful if the teaching faculty and the librarian are both committed to integrating the concepts into the course content. This case study illustrates an ideal situation: the faculty member identified the need for his students to understand the creation and use of information in remote sensing, while the librarian developed instruction to help students understand the information literacy needs for this interdisciplinary field. A collaborative relationship between the two was already in place because the librarian had been visiting the class for single-class sessions during the fall term for the past four years. As a result the librarian and faculty member were ready and willing to work together when the WVU Libraries offered faculty grants to expand the information literacy class component.

If the librarian does not have a working relationship with the science faculty, there are a number of strategies to forge one. The librarian could join professional groups with local chapters, such as the Association of Women in Science, where interaction with science faculty occurs. Regular communication with teaching and research faculty regarding trial electronic resources, new materials and new services can take the form of e-mails, postings to Twitter, and posting to a Facebook page. For smaller libraries, printing and posting recent faculty articles to a bulletin board or posting them electronically to a blog will show faculty members that the librarian is interested in their research. In addition, creating RSS feeds and alerts of faculty activity and posting these to subject LibGuides demonstrate interest.

During scheduled start-of-semester appointments with academic departmental liaisons, the science librarian can let the faculty members know about the services provided by the science librarian and the library as a whole, and can also determine the teaching faculty's expectations of the library. By introducing herself or himself, the librarian puts a face on someone ready to help with information needs. In the best case scenario, the liaison faculty members will share all information regarding library services and resources with colleagues via e-mail, departmental meetings, and other forms of communication.

Providing the very best service, helping with navigating library services and resources, and communicating openness to providing instruction will reinforce existing relationships with science faculty and potentially open new channels directly between them and the science librarian. This may occur over an extended period of time, but once the faculty members know there is a science librarian and understand his or her expertise, the relationship can grow. Such a relationship may start with a request for help in identifying obscure references or a book order and then develop into more substantive interchanges. These might include seeking suggestions for useful databases and help with using them; collection building in a specific area; discussion of open access models and other philosophical questions related to publishing; and developing strategies for teaching information literacy to science students. The first step is making the teaching faculty aware that there is a science librarian there to help them.

STRATEGIES FOR DEVELOPING A WORKING KNOWLEDGE OF THE DISCIPLINE

In addition to developing and maintaining relationships with the teaching faculty, the librarian who does not have a science background faces the challenge of not having experience with the information and bibliographic environment of the sciences. According to the 2007 ACRL Standards for Proficiencies for Instruction Librarians (ACRL, 2007), the effective teaching librarian with some subject expertise:

- Keeps current with basic precepts, theories, methodologies, and topics assigned and related subject areas and incorporates those ideas, as relevant, when planning instruction.
- Identifies core primary and secondary sources within the subject area and promotes the use of those resources through instruction.
- Uses the vocabulary for the subject and related disciplines in the classroom and when working with departmental faculty and students. (ACRL, 2007)

How does the librarian with little background or experience in the sciences get this subject expertise? It is unrealistic to believe that a librarian can become an on-the-job expert in all of the various science disciplines. However, the librarian can attempt to become more knowledgeable about the basics of the disciplines; understand the nature of bibliographic resources in the area; and become familiar with how information is gathered and shared in the discipline. Here the authors recommend some activities to help the non-science librarian gain enough subject expertise to be able to perform library instruction well. These suggestions and ideas will also aid the librarian in making contacts with faculty members.

The first recommendation is to learn about the research areas of the science departments at the institution and of the individual faculty members by examining their web pages, searching for their articles in databases, and attending presentations on their research. Take advantage of opportunities by attending campus lectures and presentations dealing with science topics and attend any vendor-sponsored webinars dealing with science topics and resources. Read current popular science books and scan the tables-of-content and articles of multidisciplinary scholarly sources such as *Science News* and *Nature* to keep abreast of current hot science topics. ACRL's Science and Technology Section email distribution list (STS-L) also provides timely information on science librarianship. Finally, take a university science class or attend a science society conference to gain find deeper insights into the information needs of the researchers in these areas.

FINAL THOUGHTS

The commitment of four class sessions to information literacy in this case study exemplifies the ideal. Given time constraints and the realization that not all teaching faculty will be willing to forego so much subject-area instruction in favor of information literacy, the minimum instruction should include how to select appropriate sources to find resources addressing a research topic and the ethical use of information. The librarian will be teaching to Standard 2 and Standard 4

The authors strongly suggest that the key to successful information literacy education involves active learning in environments where the students can integrate what they learn into their regular class work. Thus, as little as a single class period devoted to selected topics could nevertheless be very effective if the class were at a time during the semester when students can take advantage of their new skills. Students working on a term paper can immediately apply skills in database searching, and will be the most receptive to discussions about plagiarism, which are two of the most important topics for information literacy in remote sensing.

REFERENCES

ALA/ACRL/STS Task Force on Information Literacy for Science and Technology [STS-TFILST]. (2006). *Information literacy standards for science and engineering/technology.* Retrieved from http://www.ala.org/acrl/standards/infolitscitech

ACRL. (2007). *Standards for proficiencies for instruction librarians and coordinators.* Retrieved 23 Mar. 2011, from http://www.ala.org/ala/mgrps/divs/acrl/standards/profstandards.pdf

Blake, L. (2010). *Geography/Geology 455 (Remote Sensing)—LibGuides at West Virginia University.* Retrieved 18 March 2011, from http://libguides.wvu.edu/remotesensing

Blake, L., & Warner, T. (2011). Seeing the forest of information for the trees of pages: An information literacy case study in a Geography/Geology Class. *Issues in Science & Technology Librarianship, 64*(Winter). Retrieved from http://www.istl.org/11-winter/refereed2.html

Lillesand, T. M., Kiefer, R. W., & Chipman, J. W. (2008). *Remote sensing and image interpretation* (6th ed.). Hoboken, NJ: John Wiley & Sons, Inc.

Wilkinson, C. (2011). Information literacy resources for faculty and librarians—LibGuides at West Virginia University Retrieved 7 Apr. 2011, from http://libguides.wvu.edu/content.php?pid=45444&sid=335906

APPENDIX 7.1—ASSIGNMENT
Geography/Geology 455/655
Critical Reviews of Research and Scholarly Papers

Guidelines
Overview

For each week, for the next five weeks, write a short, critical evaluation of a published *research* or *scholarly* paper that deals primarily with *remote sensing* or *photo-interpretation*.

These reviews should be geared to helping you choose, and explore, your term paper.

Each review must have the following components:

1. A copy of the abstract. The abstract is the summary at the beginning of the paper. If the paper does not have an abstract, it probably is not an appropriate paper and does not fit the definition of a scholarly article as discussed in our third information literacy session.

   A *correct* bibliographic citation at the top of the page, which must follow the format of Author, (year). Title, journal (underline the name or use italics), volume and page numbers, *in that order*. If there is a journal issue, it should be in parenthesis after the volume. If you are using EndNote Web to create your reference, select the AAG Style Guide, but remember to check it carefully using this example:

   Bahria, S., N. Essoussi & M. Limam (2011) Hyperspectral data classification using geostatistics and support vector machines. *Remote Sensing Letters*, 2, 99–106.

2. A 1/2 page description of the article and any key points of interest. This should be a critical analysis in which you think about the larger issues involved.

3. A short personal evaluation paragraph where you explain how the paper relates to your interest, comment on the significance of the results, and any personal reaction you have to it.

Your review will be graded on quality of the review and your overall presentation. I expect the work to be well-edited and polished.

Language style

- Your review should use standard scientific language. Scientific language is formal, but not overly stylized or convoluted.

- Avoid colloquialisms (slang or informal speech).
- Check your spelling. Make sure each sentence is a complete sentence, and has a verb. Review the structure of your paragraphs – the ideas should flow logically. It is a good habit to proofread your work a day later, checking for mistakes. The main description of the article should be dispassionate.

Important: Plagiarism

During the second information literacy session, we discussed plagiarism and academic dishonesty. Since many journals are available on-line it is possible to actually copy directly from the paper using cut-and-paste. This is cheating. The penalties for cheating are severe. Consult the university code on cheating in the student handbook for more information. **You must use your own words throughout your review.** If you do quote, use quotation marks, followed by an appropriate citation (author, year: page number). For example:

> It has been asserted that, "high resolution imagery is particularly useful for spatial analysis, but of limited value for spectral analysis." (Jones, 2002: 438).

However, I would strongly urge you to try not to quote if possible – it is much better to use your own words. The norm in scientific scholarly papers is not to use quotations, but instead to paraphrase and summarize material. Review the slides on academic honesty in the class LibGuide if you have any questions about these points.

Geography/Geology 455/655 Term Paper Guidelines

Term paper

Each student should independently write a term paper on an aspect of remote sensing or photo-interpretation.

Topic: The first step in the paper is to turn in **a title, a one-paragraph description describing the topic of interest, and a list of no less than 4 references** on the specified date. You should have already been thinking about an appropriate narrow topic based on your work in the information literacy sessions.

Length: Approximately 3,000 words (excluding reference list). The number of references will depend greatly on the nature of the paper. A typical number might be in the 6–10 range, with 4 or 5 as an absolute minimum.

Outline &: Complete reference list. Part of presenting an argument is to develop a coherent, logical structure. Once your topic has been approved you should develop a one-page outline *that clearly shows how you will develop your presentation*. The outline will be graded, and must be sufficiently detailed that the essence of your paper is shown. Thus a list such as: Introduction, methods and conclusion, is totally inadequate. The outline needs to be sufficiently detailed that your line of reasoning is quite clear. **You need to outline the specific ideas for each paragraph, and the journal references for those particular ideas. Remember to include a complete reference list, using the appropriate format.** You may want to use EndNote Web, but be sure to check your reference list carefully.

Grading: The paper will be graded based on content and presentation. You should therefore present a logical, coherent discussion.

Plagiarism: Plagiarism is a serious offense. Cite all your sources, and be careful to use quotation marks if you use phrases or sentences that mirror those of your sources.

References: Include a list of references, using a consistent format, such as the one for the paper reviews.

In-class presentation: During the last week of classes, each student will give a 5-minute PowerPoint presentation on the term paper. You may use at most 5 PowerPoint slides. It is essential that you practice your presentation.

Some general comments

1. Aim your paper at a fellow student who has mastered the content of this course. Thus, for example, there is no need to define standard remote sensing terms or concepts such as infrared or pixel.

2. **Figures and tables** can be useful ways of presenting information. All figures and tables should be numbered, should be given titles (with a reference to their source), and should be cited in the text. (Figures are titled below the figure, tables above the figure.)

3. The strongest papers present a well-thought out overview of a topic, comparing and contrasting the different papers you read. The weakest papers tend to be summaries of individual case studies, with no clear link between the studies. If you do present a number of case studies, be sure to show the links between the studies, and have a strong concluding section drawing out common or contrasting themes.

4. **Conclusions tend to be the weakest parts of the term papers.** A strong conclusion is not simply a statement that the paper topic is an interesting or

important area. You should go back to the specifics of the topics you have covered, and make two or three general comments about the topic.

5. **References:** Provide a reference list, using the format instructions given for the paper reviews. Be sure to review how to cite (refer to) papers in the text, given in the same instructions.

APPENDIX 7.2—LESSON PLANS
Lesson Plan 1: Developing a Topic and Academic Dishonesty

Goal: This instruction session is designed to teach students how to generate a focused research question, and to understand academic dishonesty.

Objectives: Upon completion of this lesson students will:

1. Have developed a focused research topic and search statement
2. Understand plagiarism and how to avoid it
   • Demonstrate how to begin with a broad topic and narrow to a research question and a selection of keywords.
   • Demonstrate how to write a search statement
   • Explain plagiarism, it's penalties, and how to avoid plagiarism

Outline of instruction today and for other sessions
   • Session II Background on the scholarly production of knowledge
   • Session III Databases and EndNote Web
   • Session IV WoS citation searching, Internet, Cite While You Write

Standards Addressed: 1, 4
I. Generating a Research Question and Keywords from a Topic.
Explain that library research is much easier if you begin with a narrow and focused research question and as many keywords and synonyms that you can think of.
   • Show how to go from a broad research topic to a focused research question using one of the examples on the activity sheet
   • Demonstrate how to create a search statement from one of the research questions in the exercise

II. Academic Honesty and Plagiarism
   • Show slides defining plagiarism, defining University policy, and giving examples
   • Show slides of scenarios and ask students to vote "yes" by holding up a green card and "no" by holding up a red card; discuss each situation.

(Note: For larger classes, so-called "clickers", which allow students to electronically record their responses to questions, could be an effective tool for this exercise.)

III. Show LibGuide

Lesson Plan 2: Scholarly Information Production in the Geosciences

Goal: This lecture and discussion class provides students with a perspective of how information is produced and communicated in the sciences.

Objectives: Upon completion of this lesson students will:
1. Be able to identify the distinguishing features of scientific articles
2. Understand the context of scientific publications, including the relative reliability of different sources.

Standards Addressed: 1, 3

I. Why publish? The author's perspective
Explain how publishing is a professional expectation for scientists.

II. Trends in publishing
Discuss costs of journals (provide example of a journal specifically relevant to the class), including journal cost inflation, and the increasing numbers of journals, open source journals, and electronic journals.
Possible discussion topic: What does it mean to a library if it replaces its paper subscriptions with electronic subscriptions?

III. Scholarly articles within the geoscience context
This section should emphasize features that help readers identify scholarly articles. To do so, discuss distinguishing features of popular articles, and compare them to scholarly articles.
Discuss the role of conference, journal, and book publications.
Explain the typical format (structure) of scholarly articles.

IV. Peer review
Explain the peer review process, its strengths and limitations. Discuss where it is typically used, and typically not used (e.g., conference proceedings).

V. Sources of scientific information
Now from the perspective of the scholar searching for information, discuss gen-

eral sources of information (e.g., Wikipedia), news articles, trade publications, gray literature, and peer-reviewed literature.

Assignments for next meeting
Ask students to:
1. Create ILL accounts
2. Create EndNote Web accounts and download the Capture plug-in and the Cite While You Write tool

Lesson Plan 3: Effective Library Database Searching & Using EndNote Web

Goal: This instruction session is designed to teach students how to effectively search library databases and introduces EndNote Web.

Objectives: Upon completion of this lesson students will:
1. Learn to effectively search subject specific library databases and find cited articles
2. Know how to use EndNote Web to create a bibliography
- Demonstrate topic-area **databases**
- Demo how to use save references to **EndNote Web** and create a bibliography

Brief overview of library services and web site; review LibGuides
Sample topic for the session: "What trends does remote sensing show for deforestation of the Amazon?"

Standards Addressed: 2, 4

I. Databases page – EbscoHost: Agricola and Academic Search Complete
Show how to get to the databases page and the richness of resources there among the 200 databases. Discuss remote sensing as an interdisciplinary topic.

II. Sort databases by Geosciences – GeoRef
- Discuss the content of GeoRef
- Demonstrate how to use the GeoRef; (let students search their topics)

III. EndNote Web
Explain that EndNote Web is used to create a database of references for building a bibliography and for citing while writing a paper.

IV. GeoBase
- Discuss the content of GeoBase and demonstrate its search features

V. Engineering Village
- Discuss the content of Engineering Village and demonstrate its search features

Lesson Plan 4: Citation Searching, Internet Searching, and Cite While You Write

Goal: This instruction session will teach students how to effectively use citation searching and use EndNote Web for in-text citing.

Objectives: Upon completion of this lesson students will:
1. Understand the use of citation searching to review the impact of an article and to find related articles
2. Know how to use EndNote Web's Cite While You Write for citations while composing a paper
- Demonstrate how to use **Web of Science** including citation searching
- Demonstrate how to use **Cite While You Write** in EndNote Web to cite as writing

Standards Addressed: 2, 4

I. Subject searching in Web of Science and Citation searching.
Use the topics and search statements from the previous class to do a topic search in Web of Science.
- Discuss the extent and limitations of Web of Science; not the best for topic searching
- Demonstrate topic searching
- Demonstrate citation searching using a key article:

II. Using EndNote Web Cite While You Write (CWYW)

III. Students Practice
Using their research questions and research statements
1. Remind students how to create a research statement by selecting keywords and using the database cheat sheet
2. Ask students to use their own research statements based on their topics to search a database of their choosing
3. Import references to a group in EndNote Web

CHAPTER
8

IMPORTANT INFORMATION LITERACY STANDARDS FOR COMMUNITY COLLEGES

1. The information literate student determines the nature and extent of the information needed.
   2. Identifies a variety of types and formats of potential sources for information.
   3. Has a working knowledge of the literature of the field and how it is produced.

3. The information literate student critically evaluates the procured information and its sources, and as a result, decides whether or not to modify the initial query and/or seek additional sources and whether to develop a new research process.
   2. Selects information by articulating and applying criteria for evaluating both the information and its sources.

5. The information literate student understands that information literacy is an ongoing process and an important component of lifelong learning and recognizes the need to keep current regarding new developments in his or her field.
   1. Recognizes the value of ongoing assimilation and preservation of knowledge in the field.

COMMUNITY COLLEGES

Amanda Bird
Part-time Instruction & Reference Librarian
Mt. Hood Community College

Anna Montgomery Johnson
Instructor
Mt. Hood Community College

INTRODUCTION

Community college students arrive on campus with varied skills, educational backgrounds, and levels of academic preparation. Many students enroll part-time and balance work and family responsibilities with attending classes. Students in community colleges have many reasons for enrolling in courses, including recreational learning, developing skills for a specific vocation, or working toward a degree. Recently, community colleges have experienced a growth both in the number of degree-seeking students and the number of community college students who go on to earn advanced degrees (Warren, 2006). This diversity of academic goals creates a rich and varied classroom population and requires instructors to support student learning in diverse ways. Librarians delivering library instruction at community colleges are encouraged to respond to the diverse academic goals of the students when developing lesson plans and classroom activities. Information literacy for community college students is most effective when it focuses on students' goals and learning outcomes (Warren, 2006). Only by recognizing that students have a diversity of reasons for enrolling in a particular course can librarians make information literacy instruction meaningful.

Mt. Hood Community College (MHCC) is located in Gresham, Oregon and serves a population of 33,000 students from urban, suburban, and rural populations. Similar to other community colleges (Warren, 2006), MHCC students represent a cross-section of the local community and come from a wide variety of economic, cultural, linguistic, and ethnic backgrounds. To serve these varied interests, MHCC offers over 60 degree and certificate programs, including terminal 2-year degrees and transfer degrees for students continuing on to 4-year universities. In times of economic decline, community colleges see increases in enrollment as workers seek retraining opportunities (Borden, 2010), and MHCC is no exception as students from the region look for job training and educational credentials that will help them find work in a state with one of the highest unemployment rates in the nation ("Unemployment Rates," 2011).

This chapter focuses on information literacy for science courses at MHCC, specifically Biology and Fisheries Technology classes, and offers recommendations for how to enhance or evolve library instruction for science classes to be responsive to the pragmatic and vocational goals of community college students. While this chapter addresses information literacy in Biology and Fisheries Technology classes, the approaches and strategies can be extended to other science disciplines taught in community colleges. MHCC librarians work frequently with instructors in the natural resources departments to incorporate information literacy into the curriculum. Several other scientific disciplines are taught at MHCC, but the departments are smaller and tend not to schedule library instruction. MHCC's natural resources programs prepare students for employment with local and regional agencies with field-specific coursework that includes outdoor labs where students are introduced to basic principles of biology and ecology via direct exposure to wildlife in natural and hatchery habitats. While some graduates of these programs elect to continue their education at 4-year universities, most enter the workforce—typically with state or federal agencies with a focus on natural resources. The students' vocational orientation challenges librarians to ensure that information literacy instruction in these programs is directly applicable to the real-world information needs of field technicians in the natural sciences.

LIBRARY INSTRUCTION AT MT. HOOD COMMUNITY COLLEGE

At community colleges, library instruction can take many forms. Some libraries teach for-credit research skills courses and others offer drop-in workshops; some libraries may offer instruction, but it may not be coordinated to provide uniform services, and others are too understaffed to offer library instruction at all (Johnson, 2009). At MHCC, the library instruction program is robust and growing, and all librarian-led instruction is course-integrated. The MHCC library instruction program does not include tours, orientations, drop-in classes, or stand-alone research classes because MHCC librarians do not believe that these information literacy instruction delivery methods are meaningful for the career-focused student population of a community college. Because MHCC librarians only deliver library instruction that is related to a current assignment and during a regular class meeting, the research skills are always taught within the context of the student's chosen course of study. There are four major benefits to this approach:

- Students have the opportunity to experience librarian-led information literacy instruction in many different courses across the curriculum, not just once in a single required-of-everyone course.

- As students repeatedly encounter instruction librarians, they readily recognize librarians as teachers and award then the same respect as other instructors.
- As every librarian-led class is uniquely tailored to each course/instructor/assignment, the only students who can truly say "I did this already!" are those who are repeating a course with the same instructor for a second term.
- Librarians become true partners with classroom faculty in developing curriculum, often co-creating assignments and participating in assessments.

Along with many other library instruction programs at community colleges, MHCC's instruction librarians have long shared this assignment-specific information literacy philosophy; this approach continues to inform the coordination of the library instruction program, the training and mentoring of librarians involved in information literacy, and the content of the sessions.

The following example illustrates the need for the assignment specific approach. Bi102: Survey of Molecular Life and Genetics is often taught by three different instructors, and all three may arrange library instruction sessions for their classes. Each section will have a research assignment that deals with the principles of inheritance, but while one instructor might ask students to research cancer risk, another might focus on genome mapping, and the third instructor might structure his or her assignment as a debate on the ethics of genetic engineering. Because all library instruction at MHCC begins with the assignment, each Bi102 library instruction session will be unique and designed to support each instructor's interpretation of the course outcomes. MHCC's library instruction program is based on the belief that students are far more likely to perceive the librarian as an asset and a trusted resource when his or her role in their classroom is to be the guide for this assignment that is due *next Tuesday*, not as a guest lecturer about library resources they might need to use *someday*. This is true for library instruction in any setting: when the content of library instruction is as relevant as possible to the task at hand, the students will be more engaged in learning and are more likely to have positive outcomes in their coursework.

CLASSROOM ACTIVITIES SUPPORTED BY INFORMATION LITERACY STANDARDS

As instruction librarians at a comprehensive community college, MHCC librarians are challenged and rewarded by an incredibly diverse teaching environment. Within Biology courses alone, students can be first-year majors in the discipline, non-majors taking their transfer degree-required Life Science course, or terminal associate's degree students going directly into fieldwork. Students seeking to quickly enter the workforce will need to apply the knowledge they gain in the Biology course within a year or two, while students seeking transfer degrees are just beginning an educational path that may lead them to eventual careers as PhD scientists. Designing classroom activities and lesson plans that respond to the diverse information needs of community college students requires a working understanding of both the course outcomes and information literacy standards. This section will describe several strategies employed by MHCC instruction librarians in a variety of science classes and relate them to the applicable Information Literacy Standards for Science and Engineering/Technology (hereafter, Standards) (ALA/ACRL/STS Task Force on Information Literacy for Science and Technology [STS-TFILST], 2006). While these examples pertain specifically to scientific information literacy, the basic principles of information seeking and critical thinking can be readily adapted to any academic discipline in a community college setting.

Table 8.1. Selected Student Learning Outcomes and Standards Emphasized through MHCC Library Instruction in the Sciences

Desired information literacy learning outcome:	Relevant Standard and Performance Indicator (STS-TFILST, 2006):
Recognition that many types of information exist on a topic in various formats and levels.	Standard 1—Performance Indicator 2: "Identifies a variety of types and formats of potential sources for information."
A preliminary understanding of scholarly production in the discipline.	Standard 1—Performance Indicator 3: "Has a working knowledge of the literature of the field and how it is produced."
Ability to locate sources that are appropriate to the information need or task.	Standard 3—Performance Indicator 2: "Selects information by articulating and applying criteria for evaluating both the information and its sources."
Development of the information-seeking skills needed for the workplace as emphasized by future employers.	Standard 5—Performance Indicator 1: "Recognizes the value of ongoing assimilation and preservation of knowledge in the field."

Students in science classes are often assigned to choose a topic and research and evaluate the available resources. Due to the complexity of scientific topics, there are many different stakeholders, agencies, and audiences for scientific information, and each communicates in different ways and formats. One key outcome of library instruction for this type of assignment is that students gain an understanding that not all information is equal for a particular task. As emphasized by Standard 1, community college students need to develop an understanding of how various types of scientific information are produced and for what purpose (STS-TFILST, 2006). The assignment described in the Case Study (see Appendix 8.1) offers an effective method for teaching this important information. For students entering technical careers, trade-specific publications will be important resources that they will consult as part of their jobs, and the exercise helps to prepare them for research in the workplace, as well as to acquaint them with the production of scholarly information, another skill emphasized by Standard 1 (STS-TFILST, 2006). For students transferring to complete 4-year degrees, the seeds are sown for the importance of searching for peer-reviewed literature. For all students, data intended for a general audience or from non-credible sources (typical web search results) are contextualized within the continuum of other resources available on the topic.

To explore the complexity of scientific topics and resources and to teach students how to "identify a variety of types of and formats of potential sources" (STS-TFILST, 2006), the librarian selects a broad, accessible topic—for example, Columbia River salmon for a Fisheries class—and leads the class through the process of listing the various parties who are concerned with the topic and where they communicate their ideas. The librarian then asks the class to respond by generating a list of people/agencies/titles that care about the issue. These stakeholders (e.g., Native American tribes, scientists, and electric utility companies) are listed, and the librarian suggests overlooked parties as needed. Once a list of stakeholders invested in the issue is established, the librarian asks the students to consider how and where these various stakeholders communicate their ideas. This part of the exercise is typically a bit more challenging, but the end result is a detailed picture of the types of information available on the topic. Finally, the class discusses the types of information that are the most useful for the task at hand and where they can be found. Additionally, librarians can reinforce the universal importance of evaluating a source's credibility, recognizing and acknowledging bias, and currency. This classroom activity specifically fulfills the performance indicator of Standard 1 which calls on students to have "working knowledge of the literature of the field and how it is produced" (STS-TFILST, 2006), while also emphasizing the skills needed to evaluate scientific information.

Another opportunity for librarians working with lower-division science classes is to instruct students in the critical thinking emphasized by Standard 3, which states that students should be able to critically evaluate the sources and information they retrieve (STS-TFILST, 2006). Taking the time to offer instruction on critical thinking has the additional benefit of focusing on the "lifetime learning" goals emphasized in Standard 5 (STS-TFILST, 2006). Direct instruction on the critical thinking skills necessary to independently navigate the research process is always incorporated into library instruction at MHCC during instruction on various topics, including the navigation of electronic databases, selecting relevant sections from a book's table of contents, and evaluating web resources. Because the Internet will likely be the main resource students utilize after graduation, focusing on evaluating websites is often a priority for classroom faculty and librarians alike. While most MHCC students are able to use the web to conduct basic searches, they are unlikely to be experienced in critically evaluating the results. Modeling how to determine a website's credibility, how and where to find the currency of a website, and the critical thinking skills required to recognize and acknowledge bias are basics of information seeking that should not be neglected or assumed of the students—especially in a community college setting where students come from a variety of cultural and socioeconomic backgrounds and possess widely divergent technological literacy skills.

MHCC library instruction in the sciences also often involves acquainting students with the widening availability of open access journals and other science-specific web tools. Familiarizing students with these types of resources also supports a key performance indicator for Standard 5, which calls upon students to understand "emerging technologies for keeping current in the field" and further develops their ability to seek and use credible scientific literature (STS-TFILST, 2006). MHCC instruction librarians frequently introduce science students to open access journals and other digital repositories. While the scholarly level of some of these resources may surpass current information needs, the free access to these resources makes them valuable tools for future practitioners.

Working with science students who will soon enter the workplace challenges the instruction librarian to design lesson plans that go beyond the basics of navigating the library catalog and trolling databases for articles on a topic. In many courses, library instruction focuses on the search process (i.e., knowing how and where to look in the library's web site for databases, locating books on a topic, etc.); however, for students who are preparing to enter the scientific workforce as field technicians, the focus needs to be less on the search process specific to the college's library and more about how to comparatively judge the usefulness of field-specific information from a wide variety of sources. Standard 3 is also particularly important in this case.

More importantly from the student perspective, employers regularly emphasize the importance of students' ability to not only access scientific information but to also make rapid decisions about the information's usefulness. In the library instruction classroom, learning activities should complement the practical, fieldwork-minded focus of the curriculum. Instruction librarians must offer practical, effective methods for determining a source's usefulness that goes beyond the popular/scholarly dichotomy. For these soon-to-be field technicians, peer-reviewed sources are not always accessible and, more importantly, scholarly articles are not always relevant or useful to the individual's information need.

CONCLUSION

Based on the success of the MHCC library instruction program, the following recommendations can be made:

- Course-integrated library instruction enables librarians to be highly relevant, trusted resources for community college students and faculty.
- Librarians should design lesson plans for a variety of students while keeping the overall focus on critical thinking and the use and evaluation of information.
- Many community college students have vocational goals. When possible, librarians should work with department faculty to better understand the expected learning outcomes and the expectations of employers.
- Most community college science instructors are not research scientists and may lack the skills to teach their students how to find information in support of a learning goal. Librarians should recognize that their expertise is welcomed and valued by instructors and students alike.

Teaching at a community college often means working with scare resources while serving large and diverse student populations. Community college librarianship is not immune to these challenges, but librarians benefit from the college-wide focus on teaching the most critical learning behaviors for successful employment in a chosen field or continuing a course of study. In science courses when assignments are tied to a real-world workplace application, librarians are empowered to design and deliver practical library instruction lessons that are useful, relevant, and appreciated by both students and faculty.

The chapter authors would like to thank Mount Hood Community College instructors Todd Hanna, Wally Shriner, Bill Becker, and Michael Jones for sharing their ideas and their students with us.

REFERENCES

ALA/ACRL/STS Task Force on Information Literacy for Science and Technology [STS-TFILST]. (2006). *Information literacy standards for science and engineering/technology.* Retrieved from http://www.ala.org/acrl/standards/infolitscitech

Borden, V. H. (2010). Down means up. *Community College Week*, 23(8), 7–11.

Johnson, W. (2009). Developing an information literacy action plan. *Community & Junior College Libraries*, 15, 212–216.

Unemployment rates for states. (2011, June 17). *U.S. Bureau of Labor Statistics.* Retrieved from http://www.bls.gov/web/laus/laumstrk.htm

Warren, L. A. (2006). Information literacy in community colleges. *Reference & User Services Quarterly*, 45, 297–303.

APPENDIX 8.1. CASE STUDY: AN EVOLVING APPROACH TO A FISHERIES TECHNOLOGY ASSIGNMENT

MHCC instructor Todd Hanna teaches Fi101: Fishery Techniques each fall. Since 2006, Fi101 students have visited the library classroom for instruction in support of his Periodicals Review Assignment. The introduction to this assignment states:

> One of the skills employers have told us they would like you to have upon graduation is the ability to find pertinent literature related to a given topic in fisheries. You will investigate a variety of periodicals and make a determination of whether or not you would return to them in the future for fisheries-related information. All of these are skills you will use ... throughout your careers.

The assignment asks students to review articles from 18 different periodicals and to make a determination as to the usefulness of the publication when they are working in the field.

Students enter the Fisheries Technology program with a broad range of information fluency. As this assignment requires several different information literacy skills, very few students could successfully complete this assignment without librarian intervention. While the assignment has remained fundamentally the same over the years that the instructor has collaborated with librarians, the design of the library classroom activities evolved as the librarians became more familiar with the Fisheries Technology program and with the Standards (STS-TFILST, 2006).

The initial focus for library instruction sessions focused on the find aspect. If the students had to access 18 different periodicals in the MHCC Library collections, the librarian's task was to teach them how to search for and find each title. This activity routinely took the entire class time. Students left the library classroom feeling confident about accessing the MHCC Library's journal subscriptions, but they had not had time to explore the content of any single periodical. Many students struggled to complete the assignment, and most needed additional one-on-one instruction at the reference desk.

As they became more familiar with the particular learning goals of the Fisheries Technology program, MHCC librarians realized that finding is not the real goal of the assignment; rather, it is the evaluating that matters. Given that the Fisheries students receive library instruction in the fall, but by the following spring they will be working in the field for an agency that will have entirely different access points to these journals, several questions arose:

- Why spend valuable library classroom time on the "click here, then click here" steps when in 18 months that path will start at someone else's home page?

- If the intricacies of the MHCC Library's periodicals collection only matter to our Fisheries students for this single assignment, why is that the emphasis of the library instruction session?

Over the last two years, major changes were made to the library instruction for this Fisheries assignment: the librarian created a web-based handout that included a direct link to each journal and brought the print-only journals into the classroom. In the library classroom, students were taught the basics of accessing journal titles in the library, but were also shown how to access the web-based handout; the remaining class time became available for browsing print and electronic issues of the journals. Students were encouraged to explore the print-only journals while they were in the classroom, because they could access the library's online journals from home. Following these changes, all of the students leave the classroom feeling confident that they could complete this assignment on their own.

Information literacy instruction often approaches evaluating in terms of reliability, but it is equally important to teach students how to select the most useful resource from a large list. The Fi101 assignment's central task is to assess each periodical's usefulness. To determine usefulness, students must ask themselves, "When I'm working at a hatchery on the Oregon coast, is Pacific Fishing a magazine I'll want to read? Does Scientific American ever include articles about salmon?" Upon completing this assignment, Fisheries students know which journals and magazines will be worth their limited reading time when they are working, and they have learned how to efficiently and effectively browse each issue of those journals for articles that are directly relevant to their work.

ABOUT THE EDITORS

Katherine O'Clair is the Agriculture and Environmental Sciences Librarian and College Librarian for the College of Agriculture, Food and Environmental Sciences at California Polytechnic State University in San Luis Obispo. From 2004 to 2009, she served as the Life Sciences Librarian at Arizona State University in Tempe. Katherine is actively involved in ALA, ACRL, and ACRL's Science and Technology (STS) and Instruction Sections (IS). She was named a Mover & Shaker by *Library Journal* in 2007. Her professional interests include integrating information literacy into the curriculum and early career issues in librarianship. She earned her B.S. in Environmental Science from Nazareth College in Rochester, NY and her M.S. in Library and Information Studies from Florida State University.

Jeanne R. Davidson is Head of Academic Program Services at Arizona State University. She came to Arizona State University as Head of Noble Science & Engineering Library in 2007. Prior to that, she served as Physical Sciences Librarian at Oregon State University and as Science Librarian at Augustana College (IL) during which time she actively engaged in issues related to information literacy, particularly in science and engineering fields. Jeanne is actively involved in ACRL and the Science & Technology Section (STS). She shepherded the *Information Literacy Standards for Science and Engineering/Technology* through the ACRL approval process during her time as Chair of STS. She holds a B.S in Geology from Colorado State University, an M.S. in Geology/Vertebrate Paleontology from University of Wyoming and an M.L.S. from the University of Missouri.

ABOUT THE AUTHORS

Amanda Bird is a part-time instruction and reference librarian at Mt. Hood Community College in Gresham, Oregon. Amanda holds an M.L.I.S. from San Jose State University, an M.A. in American Studies from the University of Massachusetts Boston, and a B.A. in English from Lewis and Clark College. At Mt. Hood Community College, she develops and conducts assignment-specific library instruction to classes across the curriculum and loves every minute of it.

Linda Blake is the Science Librarian and Electronic Journals Coordinator at West Virginia University in Morgantown, West Virginia. She earned a B.A. (Hons.) at Glenville (WV) State College and a M.S. in Library Science at the University of Kentucky. Linda has been teaching information literacy concepts for over 30 years. Within the last five years, she has been active in promoting course integration of information literacy with emphasis on the sciences. The West Virginia University Libraries Faculty Assembly awarded her the Outstanding Librarian Award in 2010.

Elizabeth (Betsy) Hopkins has been a life sciences librarian at Brigham Young University since 2003. Currently she is liaison to Nursing; previous assignments have included Physiology, Developmental Biology, Exercise Sciences, Integrative Biology, and the Biology 100 information literacy program. She received a B.S. in Biology from Utah State University and an M.L.S. from the University of North Carolina at Chapel Hill.

Anna Montgomery Johnson is an instructor at Mt. Hood Community College in Oregon, where she teaches Business Technology and Computer Information Systems. From 2006 to 2011 she worked as a faculty librarian and coordinated the college's Library Instruction program. Anna holds an M.S. from Simmons College GSLIS and a B.A. in English from the College of William & Mary.

John Meier is a Science Librarian at the Physical and Mathematical Sciences Library in the Penn State University Libraries at University Park. His responsibilities include instruction, collection development, reference, and investigating methods of delivering library information and services. He is the librarian for the U.S. Patent and Trademark Resource Center at Penn State and liaison to the departments of Mathematics and Statistics. John holds an M.L.I.S. from the University of Pittsburgh and M.S. in Electrical and Computer Engineering from Carnegie Mellon University. His current research interests lie in using innovative technology to help library users and enhance the workflow of library professionals.

Rebecca K. Miller is the College Librarian for Science, Life Sciences, and Engineering at Virginia Tech, and she works closely with the departments of Human Nutrition, Foods, and Exercise, Engineering Education, Mathematics, and Computer Science. In addition to her direct work with these departments, Rebecca serves as the departmental Information Literacy Coordinator, develops instructional and research tools for the Virginia Tech community, provides reference services for University Libraries as a member of the SciTech Reference Team, and manages relevant library collections. Her research interests include the appropriate use of new and emerging technologies in higher education, instructional design, and scholarly communications.

Olivia Bautista Sparks is the Chemistry Librarian at Arizona State University. She earned her B.S. in Chemistry at the University of Florida; M.A. in Chemistry at the University of Arizona, and M.A. in Information Resources and Library Science also from the University of Arizona. She is interested in science information literacy and the incorporation of communication tools to assist students at their research point of need.

Timothy A. Warner is a professor of geology and geography at West Virginia University, in Morgantown, West Virginia. He holds a BSc (Hons.) from the University of Cape Town, South Africa, and a Ph.D. from Purdue University. His research specialty is remote sensing. He is co-editor of *Remote Sensing Letters* and *Progress in Physical Geography*. He has published two books: *The SAGE Handbook of Remote Sensing* and *Remote Sensing* with Idrisi Taiga. He served as a founding board member of the national remote sensing consortium, America-View. In 2007, he was Fulbright Researcher at the University of Louis Pasteur, in Strasbourg, France.

Sheila J. Young is Academic Professional Emerita of the University Libraries at Arizona State University, where she was an engineering subject specialist for engineering for nine years providing collection management, research support and information literacy instruction. Previous appointments include the University of Missouri St. Louis Ward E. Barns Library, St. Louis University Medical Library, St. Louis, where she taught information literacy to first year medical students, and the physical sciences librarian in Morgan Library at Colorado State University.